MAXIMISE YOUR MARK

resistant materials technology

revision guide

MELANIE FASCIATO
AND TRISTRAM SHEPARD

Published in 2004 by:
Nelson Thornes Ltd
Delta Place
27 Bath Road
CHELTENHAM
GL53 7TH
United Kingdom

06 07 08 / 10 9 8 7 6 5 4 3 2

A catalogue record for this book is available from the British Library

ISBN 0 7487 8996 0

Illustrations by Tristram Ariss, Hardlines, Andrew Loft and Tristram Shepard
Page make-up by Turchini Design Ltd, London
Picture research by Sue Sharp

Printed and bound in Croatia by Zrinski

Acknowledgements
Special thanks are due to Jet Mayor for his detailed comments and suggestions.

The publishers are grateful to the following for permission to reproduce photographs or other illustrative material:
Apple (UK) (p 34 bottom right)
Art Directors & TRIP Photo Library (p 15 bottom left, p 17 middle), p 18 lower middle right, p 19 upper middle, bottom right, p 30 top right, lower middle left, p 38 bottom right, p 46 all top right, p 47 top right, p 49 top right and bottom right, p 50 top right and middle left, p 51 middle left lower, p 54 middle, bottom right, p 55 top and middle lower right, p 57 top, p 59 top right, p 61 top, p 62 left)
Corel Corporation (p 15 top right, p 16 lower middle, p 40 middle left, middle right, p 43 middle right upper, p 46 bottom centre left upper, p 52 top upper left and right, middle upper, p 53 top right)
Denford (p 58 top right)
Dyson (p 30 upper middle, p 61 top right)
Getty Photo Library (p 59 left)
Hemera Technologies Inc. (p 4, p 5 top and bottom, p 6, p 7, p 8, p 9, p 15 middle left and centre, p 16 middle and bottom, p 17 top, lower middle left, and bottom right, p 18 all except lower middle right, p 19 middle left top, p 20 upper middle right, bottom left, bottom right, p 21 bottom left, p 22 middle right lower, p 23 middle right, bottom left, p 25 all, p 26 lower middle, p 27 middle left and right, p 28 all, p 29 all, p 30 lower middle right, bottom right, p 31 all, p 32 all, p 33 all, p 34 all except bottom right, p 35 all, p 36 all, p 37 all, p 38 middle left and right, p 39 top right, lower middle right, bottom right, all except middle left, p 41 top right middle left lower, bottom left and right, p 42 all, p 43 all except middle right upper, p 44 all except middle left, p 45 all, p 46 middle left, bottom centre, p 47 bottom right, p 46 all, p 49 middle right and bottom left, p 50 middle right, p 51 top right, bottom right, p 52 middle toplower left, middle lower, bottom right, p 53 middle left, bottom left and right, p 54 bottom left, p 55 middle right upper, p 56 all, p 57 left middle, bottom left, p 58, middle left and bottom right, p 59 middle right, bottom right, p 60 all, p 61 bottom right, p 62 bottom, p 64 lower middle left, centre and right)
Martyn F Chillmaid (p 17 middle, p 19 middle left bottom, p 21 bottom right, p 22 bottom and top right, p 23 top, top left, middle left, far left, p 24 all, p 26 bottom left, p 41 middle left, p 51 middle right, p 52 bottom centre)
Nova Development Corporation, p 20 top right, lower middle right, p 21 top right, p 22 middle left, p 26 top and middle right, p 38 top right, p 46 bottom left centre lower, p 47 middle left lower, p 50 bottom left, p 52 lower top right)
Tristram Shepard (p 27 bottom right)
Wacom Technology (p 41 lower top right)

Every effort has been made to contact copyright holders. The publishers apologise to anyone whose rights have been inadvertently overlooked, and will be happy to rectify any errors or omissions.

Contents

Introduction

Maximise Your Mark has been developed to help make revision for the GCSE written paper more worthwhile, and more enjoyable. It provides a comprehensive, structured programme of study and revision that helps you identify what you know and don't know. The first section also gives you tips on the best ways to tackle questions.

There are a number of ways in which you might use this book:

- as a weekly programme of study during the final year of your GCSE course
- as a revision checklist to work through in the months just before you take the written paper.

The book is divided up into ten topics that cover the examination specification. Each topic has five sections. Each section is contained on a single page and includes an example question for you to try. Most sections also include a handy cross-reference to pages in the GCSE *Design & Make It! Resistant Materials Technology* textbook, where you will find more details. Keywords and relevant web-links are provided, where appropriate.

A CD-Rom version of this book may be available in your school. Ask your teacher if it is on your school's network, or on any stand-alone PCs, and if there are copies of the CDs you can purchase for home use. The electronic version also includes animated illustrations, multiple-choice questions, sample answers with examiners' comments and a full glossary.

The Design & Make It! team hope that *Maximise Your Mark: Resistant Materials Technology Revised Guide* will help you get the marks you deserve in your written paper.

Good luck in your exams!

Tristram Shepard
Series editor

Improve Your Technique

Did you know that doing well in your Design & Technology written paper involves more than just knowing all the information contained in this study guide? You also need to know how to tackle the particular types of questions you will be asked – what the examiner is going to be looking for, how long to spend on each answer, how to approach sketching, etc.

The study and revision techniques you are learning in your other subjects are going to be very useful with D&T too, but the D&T written papers are a little bit different from most of the others that you sit. The next few pages are intended to help focus your revision on exactly what's required for the Resistant Materials examination.

Top Tips

Here are six top tips to help you maximise your mark:

1. Read the question very carefully
2. Look closely at the available marks for each part of the question
3. Use the marks as a guide to how long you should spend on each question
4. Use Resistant Materials keywords and phrases whenever possible
5. Use clear, neat rough sketches
6. Remember to use your coursework

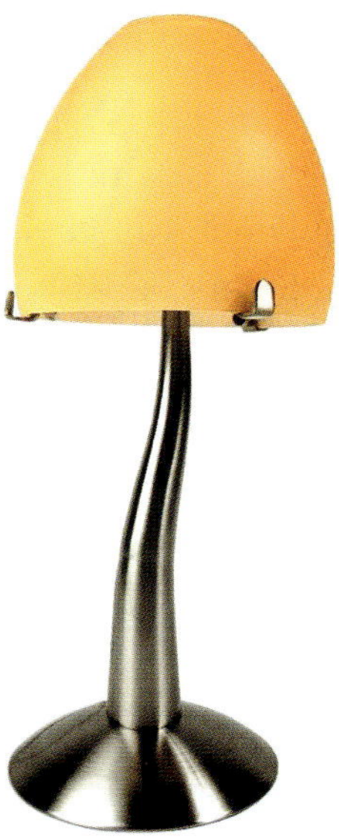

1 Read the question very carefully

Sounds obvious doesn't it? However, many students throw marks away simply because they've not paused for a moment and checked what's really being asked for. It's all too easy to spot a couple of familiar words in a question and start writing everything you know about them. No D&T question ever works like this!

Each question is asking you to show and apply your knowledge in some way. For example, you may need to:

- discuss the advantages and disadvantages of...
- provide a number of examples of...
- explain what is meant by a specific term when used in a particular situation...

- Advantages and disadvantages
- Analyse
- Calculate
- Compare (and contrast)
- Describe
- Detail
- Develop
- Differences (and similarities)
- Draw
- Evaluate
- Give reasons
- Identify
- Label
- List
- Name
- Outline
- State
- Study
- Use notes and sketches

On the left are some other words commonly used in questions – you need to be clear on what each is instructing you to do.

When the examiner comes to mark your paper he or she has to follow a set mark scheme that matches the specific requirements of each question. So if there are two marks for giving two advantages and two marks for giving two disadvantages, you can only score a maximum of half-marks if you only discuss the advantages – however many you cover. In a similar way, you'll lose marks if you fail to provide the correct number of examples required.

2 Look closely at the available marks for each part of the question

To make life a bit easier, the D&T written papers give you plenty of information about how many marks are available for each part of each question. In some cases it might even show you that there are, say, two marks for stating the advantages and two for the disadvantages. In another it might be a little less obvious, but it's usually quite easy to work out what the marking scheme is going to be. For example, what do you think the marking scheme for the following question is likely to be?

> Explain two differences between thermosetting plastics and thermoplastics, giving an example of each. *(6 marks)*

Here you are being asked for two examples, which are likely to be allocated a mark each. That leaves four marks for the explanation of the two differences, i.e. two marks for each difference. These two marks are likely to be subdivided, with one mark for identifying a suitable difference, and the other for the quality of explanation you provide.

You need to get into the habit of providing examples and explanations. Unless the question asks for a list, it's unlikely that a simple one word answer is going to be enough to get you the marks. Basically for each section you study or revise you can pre-prepare a number of examples or explanations, and then use whichever supports the answer to the question most effectively.

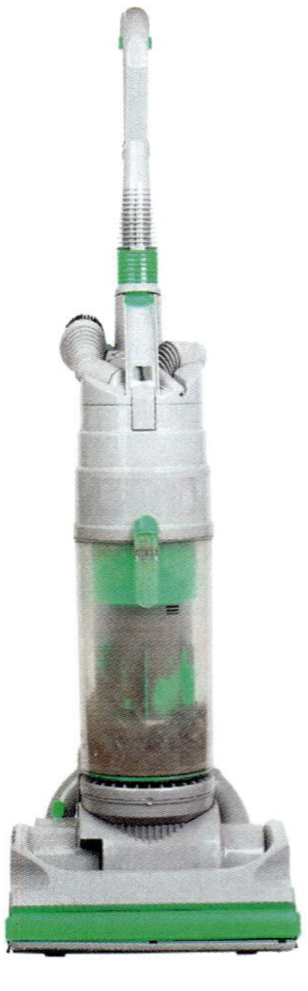

During your course you should get to know the names of some designers and things that have been designed. Try to use specific examples in which you name the names of the designer, product model and/or manufacturer. So instead of just saying an 'MP3 player' you might write 'the Apple MP3 iPod, designed by Jonathan Ive', or 'the Cyclone cleaner, designed by James Dyson', instead of just 'a vacuum cleaner'. Examples like these will quickly earn the marks.

3 Use the marks as a guide to how long you should spend on each question

In the D&T written papers there's a very simple guide as to how long to spend on each question – a mark a minute! So if a question is worth six marks you should be spending around five or six minutes attempting to answer it. If it only takes you, say, one minute, then it's possible that you haven't provided enough detail. If the six minutes is nearly up and you're only halfway through, then you're probably writing far too much.

Remember that you'll need to include some initial time to read the question and think about your answer – don't just start writing. Start by checking the mark scheme, working out what examples you are going to use, thinking of the appropriate technical terms, etc.

Getting the timing right on longer questions that ask you to sketch and develop ideas is rather more difficult. It's very important that you practise answering these sorts of questions in the appropriate amount of time before the final examination. Again use the breakdown of the amount of marks as a guide. If there are, say, 15 marks to sketch three ideas, and 20 marks to develop one in more detail, make sure you spend about 15 minutes on the initial sketches and 20 minutes on the more detailed drawing. There is more advice on sketching on page 8.

With D&T papers you are advised to attempt the answer the questions in the order in which they are set. This is because the answer to one may lead on to the answer to a following question. However, it's still important not to spend too much time on a question you are finding difficult. Do what you can in the given time, and then move on. If there's time left at the end you can then go back and try to finish it off.

4 Use Resistant Materials keywords and phrases whenever possible

On each page of the study guide that follow you'll find a number of keywords. You need to know what each one means and how to use it correctly in a sentence. You will find that you already know what many of them mean, which is just as well as in all there are about two hundred! Whenever you come across a word or phrase you are not familiar with, make a point of finding out what it means – ask your teacher or look it up in a textbook. If your school has the *Maximise Your Mark: Resistant Materials Technology* CD-Rom you can check the glossary, or use the QuickSearch option.

Aim to use as many of these keyword terms as possible in your written paper: they really help show that you know what you're talking about. In many questions each one you use might easily earn you an extra mark, and make the difference between one grade and another.

It's also a good idea to try and remember the group of keywords that apply to each section. When you come across a question that relates to that section, thinking of as many keywords as possible will help you recall what it's all about. For example, here's a group of keywords:

- ICT (Information and Communication Technology)
- CAD (Computer Aided Design)
- CAM (Computer Aided Manufacture)
- CNC (Computer Numerical Control)

Suppose a question comes up about the use of CAD. If you have learnt the whole group of keywords it will help you quickly recall a much wider range of information about CAD, and you might also be able to use some of the other keywords in your answer.

5 Use clear, neat rough sketches

Despite being called a 'written' paper, you will probably find you have to do some sketching. In fact you may find a series of related questions that ask you to generate and develop a design idea using sketches. These questions might take you nearly an hour, and be worth half the marks of the whole paper. Learning how to tackle these sorts of questions is therefore essential if you are going to get a good grade: make sure you do plenty of practice questions like these before the final paper.

In the examination you are not expected to do measured drawings, i.e. using a ruler. This is mainly because it would take too long. In doing your coursework you should have developed the ability to use clear, neat rough sketches – and these are exactly what are required here. Remember to add notes to your drawings to help explain your thoughts – these might help explain your thinking, e.g. 'This shape will be easy to vacuum form', or may be problems you are considering, e.g. ' What would be the best way to join these pieces together?' You can also use words to help describe a shape or form you are finding difficult to sketch.

Avoid adding colour to your sketches (unless of course the question specifically asks you to do so), and only use it sparingly for emphasis.

As always, make sure you read the question carefully and are covering exactly what is asked for in terms of the number of drawings, whether notes and evaluations are needed, or particular types of drawings, e.g. plans, elevations, 3D, etc. If it's up to you, then choose 2D views as they are easier and quicker to complete.

On the next page are examples of a very successful response to a series of shorter questions, worth about half the total marks for the paper! Could you do something similar in the exam in about an hour?

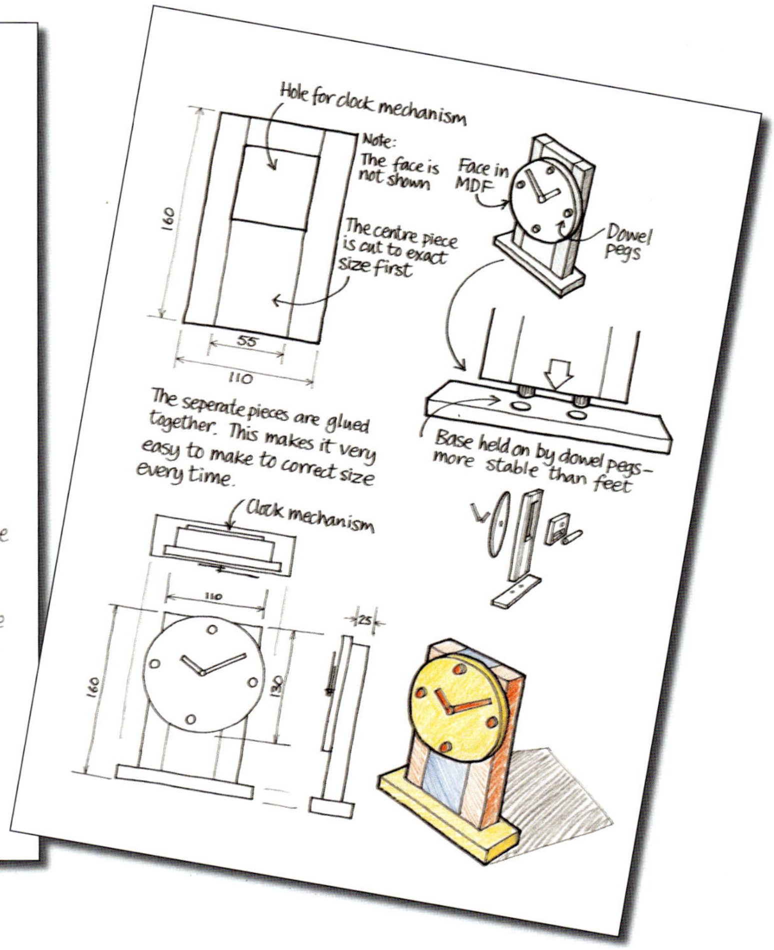

6 Remember to use your coursework

Sometimes questions ask you to refer to your GCSE coursework, perhaps as an example. Be aware that the person marking your written paper will not have seen your practical work or design folder, so you will probably need to give some brief information about what you designed and made.

You may find it helpful to think back to your coursework. For example if you need to give an example of something designed for batch production, or maybe to explain the process of vacuum-forming, the work you did during your course might help remind you of the key points.

Other questions might show you a photograph of a product and ask you to develop its specification, or undertake an analysis and evaluation of it. You will have probably done similar exercises during your course. Although the product might be slightly different, think back to the sort of observations you made, and see if they apply to this example.

Finally, remember that the examination is trying to find out what you *do* know and understand, not what you don't. We all hope you manage to Maximise Your Mark!

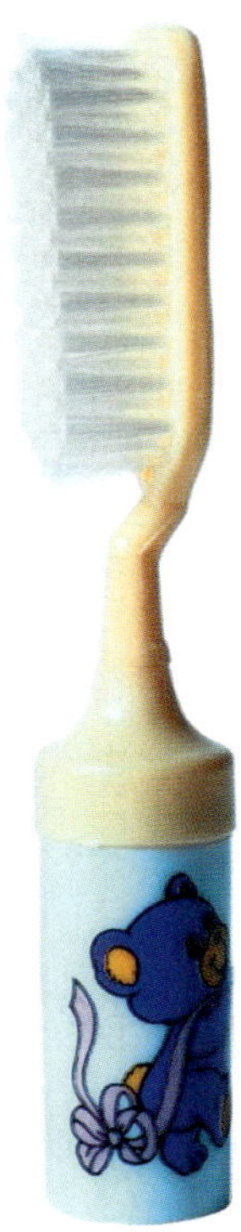

MAXIMISE YOUR MARK

Resistant Materials Technology

Revision Planner

Name ____________ Form ____________

Resistant Materials Technology	Topic / Section Start Date	Target Finish Date	Actual Finish Date	How well did I do on the written paper question? How could I have maximised my marks?	What do I need to revise again for this section?
Topic 1 Properties and Characteristics of Materials					End of Topic Review ✓
Section 1 General properties					
Section 2 The properties of metals					
Section 3 The properties of plastics					
Section 4 The properties of wood					
Section 5 The use of new materials					
Topic 2 Components and Adhesives					End of Topic Review ✓
Section 1 Nuts, bolts and washers					
Section 2 Wood screws					
Section 3 Nails, pins and rivets					
Section 4 Fixings and fittings					
Section 5 Adhesives					

MAXIMISE YOUR MARK

Revision Planner

Name ____ Form ____

Resistant Materials Technology	Topic / Section Start Date	Target Finish Date	Actual Finish Date	How well did I do on the written paper question? How could I have maximised my marks?	What do I need to revise again for this section?
Topic 3 Product Analysis and Evaluation				End of Topic Review ✓	
Section 1 The analysis of designs and products					
Section 2 Aesthetics and function					
Section 3 Past and present products and designers					
Section 4 Sketching products in 2D and 3D					
Section 5 Design specification criteria					
Section 6 Testing and evaluation					
Topic 4 Social, Cultural, Moral and Environmental Issues				End of Topic Review ✓	
Section 1 Social, cultural and moral influences					
Section 2 Ergonomics					
Section 3 Anthropometrics					
Section 4 Consumer rights					
Section 5 Environmental issues					

Revision Planner — Name ______ Form ______

MAXIMISE YOUR MARK Resistant Materials Technology	Topic / Section Start Date	Target Finish Date	Actual Finish Date	How well did I do on the written paper question? How could I have maximised my marks?	What do I need to revise again for this section?
Topic 5 Health and Safety				End of Topic Review ✓	
Section 1 Identifying hazards and reducing risks					
Section 2 Using workshop tools safely					
Section 3 Creating a safe working environment					
Section 4 Ensuring that end products are safe for consumer use					
Topic 6 Techniques and Processes (1)				End of Topic Review ✓	
Section 1 Holding devices					
Section 2 Marking out tools					
Section 3 Cutting tools					
Section 4 Drilling tools					
Section 5 Files and abrasive papers					
Section 6 Various finishes					

MAXIMISE YOUR MARK

Revision Planner

Name ______ Form ______

Resistant Materials Technology	Topic / Section Start Date	Target Finish Date	Actual Finish Date	How well did I do on the written paper question? How could I have maximised my marks?	What do I need to revise again for this section?
Topic 7 Techniques and Processes (2)				End of Topic Review ✓	
Section 1 Manufacturing in metal					
Section 2 Manufacturing in plastics (1)					
Section 3 Manufacturing in plastics (2)					
Section 4 Manufacturing in wood					
Section 5 Machine tools and techniques					
Section 6 Production process systems					
Topic 8 Systems and Control				End of Topic Review ✓	
Section 1 Structural Systems					
Section 2 Mechanical Systems					
Section 3 Production Systems					
Section 4 Quality Control Systems					

MAXIMISE YOUR MARK

Revision Planner

Name ______ Form ______

Resistant Materials Technology	Topic / Section Start Date	Target Finish Date	Actual Finish Date	How well did I do on the written paper question? How could I have maximised my marks?	What do I need to revise again for this section?
Topic 9 Information and Communication Technology				End of Topic Review ✓	
Section 1 How to use a wide range of software when designing					
Section 2 How CAD/CAM systems are set up					
Section 3 How to use 2D and 3D CAD software					
Section 4 The advantages and disadvantages of using CAD/CAM					
Topic 10 Industrial Practices				End of Topic Review ✓	
Section 1 The client, designer, manufacturer and user					
Section 2 Industrial production planning and costing					
Section 3 Advertising and marketing					
Section 4 Product safety symbols					
Section 5 Drawing conventions					

Here's what you need to know...

about the general properties of resistant materials.

KEYWORDS
Do you know what the following terms mean?
- Properties
- Working characteristics

WWW.
www.matweb.com/search/SearchSubcat.asp

What are Properties and Working Characteristics?

When designing a product using resistant materials, the properties and working characteristics of the materials are very important.
The designer needs to know about their:

- **properties** – how materials perform in everyday use.
- **working characteristics** – cutting, joining and finishing them.

This section is mainly concerned with the properties of materials. The working characteristics of materials are covered in other sections and topics.

Common Properties

There are eight common properties for all resistant materials:

1 Stiffness
Stiff materials are not **flexible**; they cannot be squeezed or stretched easily.

2 Strength
Strong materials are not easy to break or deform.

3 Toughness
Tough materials can **absorb** a lot of impact before they break.

4 Thermal conductivity
Thermal conductivity is how easily heat is passed through a material.

5 Electrical conductivity
Electrical conductivity is how easily electricity is passed through a material.

6 Hardness
This is how resistant to wear the material is.

7 Environmental resistance
This is how easily the material rots or corrodes in sunshine and rain.

8 Density
This refers to how heavy a material is when compared to its volume.

Outdoor furniture is often made from teak, because this wood is resistant to rot.

Property Sharing

Metals, woods and plastics all share these eight properties. However, different materials display different combinations of these properties. For example, metal is tougher than plastic, but plastic has a better environmental resistance than wood.

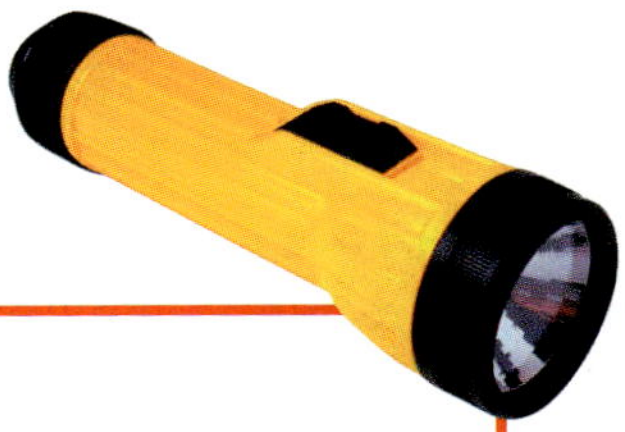

The casing of a torch is usually plastic or rubber, both of which have low electrical conductivity.

Sometimes a design requires a material that is high in one or more particular property, such as toughness. However, a design might also require a material that is poor at doing something, such as electrical conductivity.

This table shows the ways in which metal, wood and plastic vary in their properties.

Property	Metal	Wood	Plastic
Stiffness	High	Medium	Low
Strength	High	Medium	Low
Toughness	High	High/medium	Medium/low
Thermal conductivity	High	Low	Medium
Electrical conductivity	High	Low	Low
Hardness	High/medium	Medium/low	Low
Environmental resistance	High/medium	Medium/low	High
Density	High	Low	Low

There are also differences between different types of the same materials. For example, some metals are stronger than others and some plastics are stiffer than others.

Written Question

Answer the following question on lined paper. Do not spend more than 6 minutes writing your answer.

Study the photograph of the kettle below.

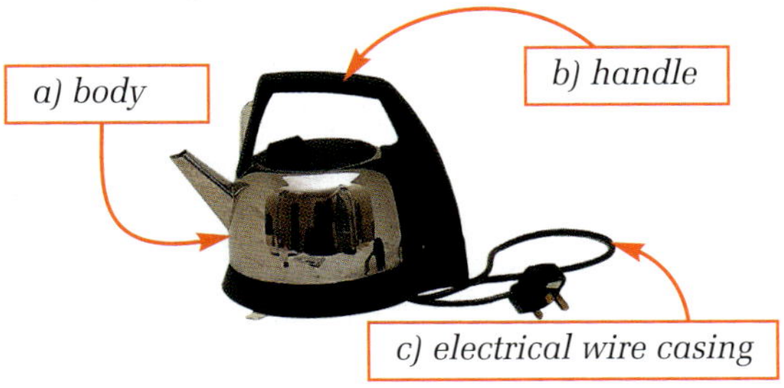

Identify an appropriate type of material for each part of the labelled kettle. Give a reason for each choice with reference to a suitable property of the chosen material. *(6 marks)*

Here's what you need to know...

about the properties of metals.

See *Design & Make It! Resistant Materials Technology* Revised pages 54–55, 58–59, 62–65, 69, 130–131 (46–47, 50–51, 54–57, 61, 122–123 earlier edition).

KEYWORDS

Do you know what the following terms mean?

- Ferrous
- Non-ferrous
- Alloy
- Annealing

www.

www.azom.com/materials.asp
www.azom.com/Applications.asp
www.designandtech.com/ (go to Resistant Materials)

Metal Matters

Designers make choices about which metals they are going to use by considering the properties of the metal and whether these suit the needs of their design. For example:

- Does it need to be light or heavy?
- Does it need to be malleable, flexible or rigid?
- Does it need to resist corrosion?
- What colour is it?
- Is it shiny or dull?
- Is it tough?
- Is it environmentally friendly?
- Is it a good conductor of heat?
- Is it a good conductor of electricity?
- Would an alloy of metals provide a good solution?
- How much does it cost?
- How can it be finished?
- How can it be joined?
- What manufacturing processes are needed to shape it?

Ferrous or Non-ferrous?

Metals are categorised as **ferrous** or **non-ferrous**. This refers to whether they contain iron or not. For example, mild steel is a ferrous metal – it contains iron and 0.15% to 0.3% carbon. Adding **carbon** gives it different properties from pure iron, making it tough and malleable.

Aluminium is a non-ferrous metal – it does not contain any iron. It is made from the pure metal with no additions.

Alloys

Alloys are combinations of metals. For example, brass is made from 65% copper and 35% zinc. The resulting alloy combines the properties of both metals.

Heat Treatment

When metal is heated and left to cool slowly the process is called **annealing** which makes it more malleable and ductile. An **oxide** is formed on the surface but can usually be removed by soaking in alum solution or pickling.

Written Question

Answer the following question on lined paper.
Do not spend more than 5 minutes writing your answer.

A designer is working on an idea that combines metal and wood in a design for a chair.

i) Which metal would be most suitable for constructing the chair legs? *(1 mark)*

ii) Write about **four** properties that this metal has which makes it suitable for this use. *(4 marks)*

Here's what you need to know...

about the properties of plastics.

See *Design & Make It! Resistant Materials Technology* Revised pages 32–35, 40–41, 60–61, 63, 132–133 (24–27, 32–33, 52–53, 55, 124–125 earlier edition).

KEYWORDS

Do you know what the following terms mean?

- Thermoplastics
- Thermosetting plastics
- Plastic memory
- Laminated

www.

www.azom.com/materials.asp
www.azom.com/Applications.asp
www.designandtech.com/ (click on Subject Search, then Resistant Materials)

Plastic Properties

Designers make choices about which plastics they are going to use by considering their various properties and whether these suit the needs of the design.
For example:

- Does it need to be light or heavy?
- Does it need to be flexible or rigid?
- Does it absorb sudden impact?
- Can it be coloured?
- Is it opaque or translucent?
- Is it a good chemical insulator?
- Is it a good insulator of heat?
- Is it a good insulator of electricity?
- Is it environmentally friendly?
- Is it tough?
- Is it durable?
- How much does it cost?
- How can it be joined?
- What manufacturing processes are needed to shape it?

Plastics are categorised as **thermoplastics** or **thermosetting** plastics. This is a good guide to the **working characteristics** of the plastic.

Thermoplastics

Thermoplastics are shaped when they are hot. They are less resistant to heat and fire than thermosetting plastics. Thermoplastics can be re-heated after being formed. They usually return to their original flat sheet form. This is known as **plastic memory**.

An example of a thermoplastic often found in school is acrylic, one trade name for this is Perspex™.

Thermosetting Plastics

Thermosetting plastics are stronger, harder and stiffer than thermoplastics. They are resistant to heat and fire. A chemical reaction is involved in their forming.

Polyester resin is an example of a thermosetting plastic. It is **laminated** with glass or carbon fibres for moulding GRP (glass reinforced plastic) car and boat bodies.

Written question

Answer the following question on lined paper.
Do not spend more than 8 minutes writing your answer.

A manufacturer wants to produce a new range of kitchen bins. He has approached you, as a designer, with the idea of using high-density polythene.

What four properties does this plastic have that make it an ideal choice? Give your reason for each choice. *(8 marks)*

Here's what you need to know...

about the properties of wood.

See *Design & Make It! Resistant Materials Technology* Revised pages 82–83, 100–105, 134–135 (74–75, 92–97, 126–127 earlier edition).

KEYWORDS

Do you know what the following terms mean?

- Natural wood
- Manufactured board
- Softwood
- Hardwood
- Medium Density Fibreboard (MDF)
- Plywood
- Lamination

www.

www.azom.com/materials.asp

www.azom.com/Applications.asp

www.designandtech.com/ (click on Subject Search, then Resistant Materials)

Wood – You Know It!

There are two categories of material that are often referred to as wood: **natural wood** and manufactured board. Designers make choices about which natural wood or **manufactured board** they are going to use by considering whether the properties suit the needs of their design. For example:

- What colour is it?
- Can it be veneered or laminated?
- Can it be bent in a curve?
- Is it environmentally friendly?
- How much does it cost?
- How can it be finished?
- How can it be joined?
- Is it easily worked?

- Does it need to be light or heavy?
- Does it need to be straight grained?
- Does it need to be free from knots?
- Does it need to be weatherproof?
- Does it come in wide boards?

Natural Boards

Natural wood, or timber, is categorised as either **softwood** or **hardwood**. However, this does not describe the working characteristics of the wood – softwood is not necessarily soft, and hardwood is not necessarily hard.

Softwood comes from trees that keep their leaves in winter. These are usually **conifers**, for example, Pirana Pine.

Hardwood comes from **deciduous** trees. These trees lose their leaves in the winter, for example, Beech.

Manufactured Boards

Manufactured boards are processed from particles, sheets or small pieces of wood. For example, to make **Medium Density Fibreboard (MDF)** small particles of wood are mixed with glue, heated and compressed into flat sheets.

To make **plywood**, veneers or thin sheets of wood are glued together, heated and pressed. This is also known as **lamination**. Manufactured boards are very stable.

Written question

Answer the following question on lined paper.
Do not spend more than 8 minutes writing your answer.

A designer is producing a specification for a batch of wooden toy-boxes to be made for young children. A decision is needed as to whether they should be made from redwood or medium density fibreboard.

i) Explain three properties for each material that makes them suitable for this purpose. *(6 marks)*

ii) Which of the two materials would you choose and why? *(2 marks)*

Here's what you need to know...

about the properties of new materials.

KEYWORDS

Do you know what the following terms mean?

- Composite materials
- Smart materials
- Motion control gels/gases
- Shape Memory Alloys (SMAs)

www.

www.designcouncil.org.uk
www.tep.org.uk
www.laboratorytalk.com
www.designinsite.dk

New Materials

A range of new materials has been developed over the past twenty years. There are two main types of new materials – **composite materials** and **smart materials**.

Composite Materials

Metals, woods and plastics all have particular properties, which are 'set'. Now, new materials can be made to have a particular property that a designer wants. This means that the designer can **specify** the properties that are required for the job, rather than be restricted by the existing properties of standard metals, woods and plastics.

These new materials are usually composites. This means they are created from different mixtures of other materials and chemicals that combine together to give the qualities that are required. For example, conductive foam is used to protect integrated circuits when they are being transported. Their legs are inserted into the block of conductive foam. 'Normal' foam would protect the circuits from knocks and bumps during transportation, however, conductive foam has been designed to have properties that stop the build up of static electricity – as well as protecting the integrated circuits from damage.

Smart Materials

These materials are referred to as 'intelligent' or smart materials. The properties of smart materials change in response to a change in their environment, for example, temperature and light. Some familiar examples of these materials are the plastics used for sunglasses that get darker when used in bright sunlight and ceramics and plastics that change colour when heated.

Many applications of smart materials include **electrical components** embedded within the material. These can be connected to computer systems. Other ways of describing smart materials are **active**, **adaptive** or **sensing materials**. They are often created to perform specific tasks in response to particular situations.

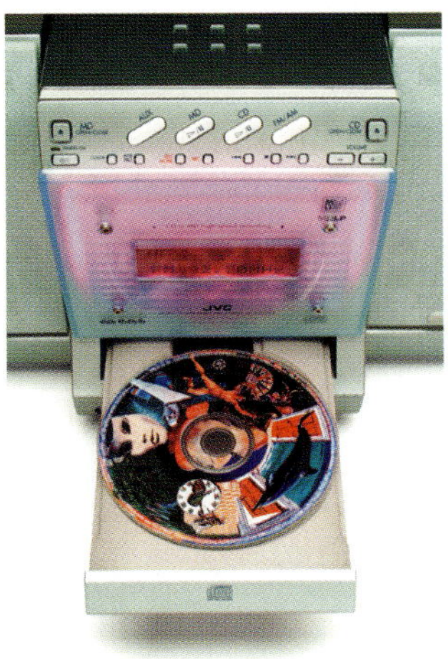

Motion Control Gels and Gases

Motion control gels and **gases**, for example, 'smart grease', are used to regulate the movement of components that are in contact with each other. Some examples are sliding microscope barrels, variable resistors and slow spring-return CD drawers.

Shape Memory Alloys (SMAs)

Shape Memory Alloys change shape and volume in response to certain conditions, for example, temperature changes. They can be 'trained' to reverse their transformation, returning to their original shape and size. These are often used in wire form to make springs and interwoven with fibres to make fabrics that change shape and size according to temperature. Bendable spectacle frames are made from Shape Memory Alloys.

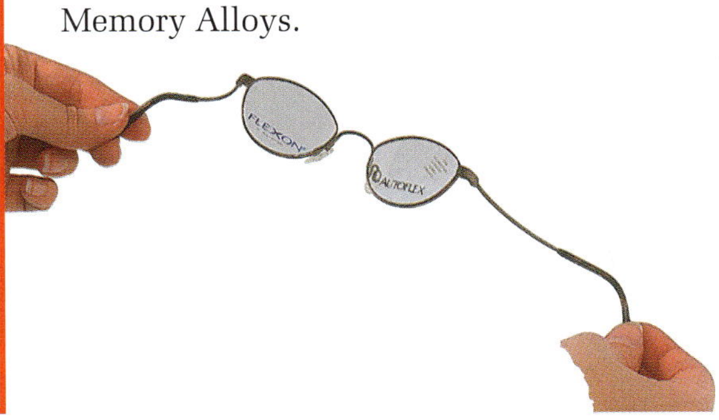

Written question

Answer the following question on lined paper. Do not spend more than 5 minutes writing your answer.

Explain the difference between a composite material and a smart material. Give a specific example of each. *(5 marks)*

Here's what you need to know...

about nuts, bolts and washers.

See *Design & Make It! Resistant Materials Technology* Revised pages 120–121 (112–113 earlier edition).

KEYWORDS

Do you know what the following terms mean?

- Pre-made components
- Nuts and bolts
- Washers

Fixing It All Together

Deciding on the way in which different pieces of materials are joined together is an important part of the design process. There are a variety of **pre-made components**, such as screws and nails, and different adhesives that are available. Durability, safety, cost and ease of working all need to be taken into consideration. Ideally, there should be as few fastening operations as possible in order to reduce production time and costs.

Nuts and Bolts

Nuts and bolts are removable fastenings for metal and wood. A bolt has a screw thread. The nut fits on to the end of the bolt and it revolves up the screw thread towards the head. Bolts are usually used with a washer that is placed underneath the nut between the nut and the screw thread.

Types of nuts

There are a range of types of nuts to choose from.

- **Hexagon**: the basic nut, tightened with a spanner.
- **Wing nut**: used with removable fittings, tightened by hand.
- **Locknut 1**: used with a standard nut to stop the first nut from coming loose. The two nuts are tightened against each other.
- **Locknut 2**: nylock type nuts have a nylon insert and resist coming loose without a second nut being used.
- **Cap nut**: this gives a decorative finish at the end of a bolt.

Bolts

There are two main types of bolts to choose from.

- A **standard bolt**. This is only threaded for part of its length.
- **Set machine screws**. These are bolts that are threaded for their whole length. They are screwed into a threaded hole to fix two parts together.

Both types come in a variety of lengths and diameters.

There are many types of bolt heads, including slot head, countersink, and cheese head. Hexagonal or socket head screws are known as Allen screws.

Sizes of Bolts

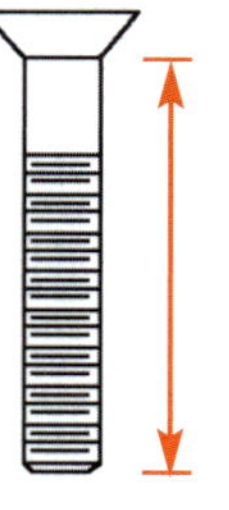

Bolts are measured by the length of the body, not including the head.

'M' means the outside thread diameter in mm. The thread can be either coarse or fine.

Washers

Washers are available in a range of sizes and thicknesses. Their size is the diameter of the hole measured in mm. There are several types of washer:

- **Plain washer**: used to spread pressure over a larger area and protect the surface of the part being joined.
- **Spring lock washer**: stops the nut from coming loose. Made from spring steel.
- **Toothed washers**: these have spikes that bite into soft material, providing extra grip.

Written Question

Spend about 6 minutes answering the following question. You will need some paper and something to write with.

Here is a list of products where a nut and bolt would be a suitable method of fixing.

1 A desk light where the arm can be adjusted to different angles.
2 A climbing frame in a children's play area.
3 A flat-pack bookcase.

For each of the three products above give details of the types of nuts, bolts and washers (if required) that you would specify for use in each product. Give a reason for each choice.

(6 marks)

Here's what you need to know...

about different types of screws.

See *Design and Make It! Resistant Materials Technology* Revised pages 101 and 120–121 (93 and 112–113 earlier edition).

KEYWORDS

Do you know what the following terms mean?

- Screws
- Shank
- Gauge
- Clearance holes
- Pilot hole

Fastening Wood

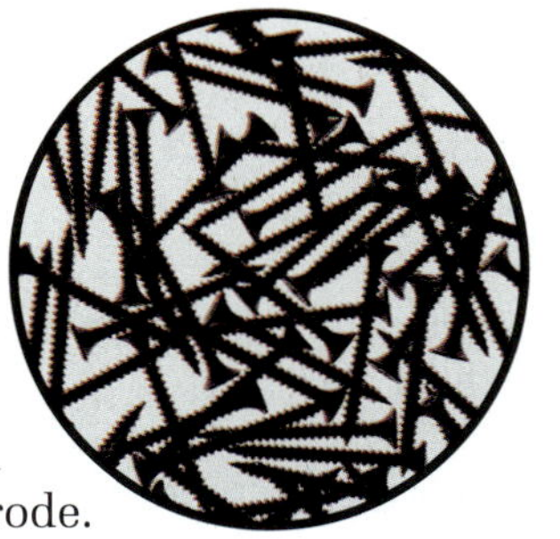

Wood is often joined using **screws** made from steel, brass, aluminium and stainless steel. They are available in a variety of different finishes including: bright zinc plate, nickel plate, chrome plate, brass plate and black japanned.

- The heads of **steel screws** should be painted with metal primer if they are used in damp conditions, as they corrode. Non-corroding steel screws are available.
- **Brass screws** are used with decorative brass fittings. They are expensive and break easily. It is important that the correct size of pilot hole is drilled first.
- **Aluminium screws** are useful for wood as they do not stain or rust.

Pieces of wood can also be jointed together without the need for fastenings or adhesives, though these are sometimes used as well to give extra strength. However, using joints can add a lot of time to the construction process. There is more information about using joints in Topic 7 section 4.

Sizes of Screws

The size of a screw is the diameter of the **shank**, or unthreaded part. Screw size is measured by **gauge** numbers. These go from 4 to 12 in even numbers. 4 is small and 12 is large.

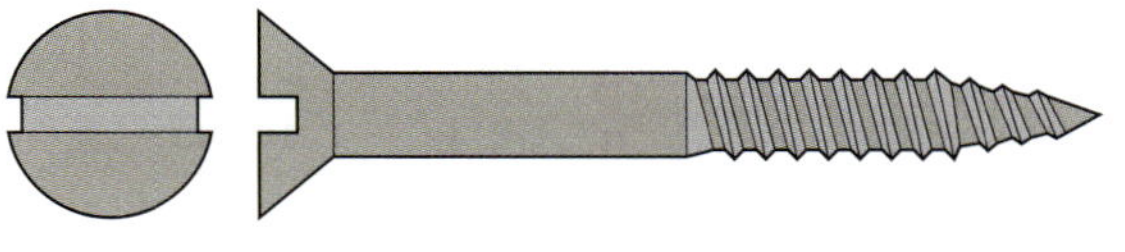

The length of a screw is measured from beneath the head to the tip. Each gauge of screw comes in a range of lengths, varying from 10mm to 100mm.

Clearance Holes and Pilot Holes

Clearance holes are drilled in the top piece of wood and need to be big enough to allow the shank of the screw to pass snugly through it. The **pilot hole** is drilled into the lower piece of wood and needs to be smaller than the threaded part of the screw.

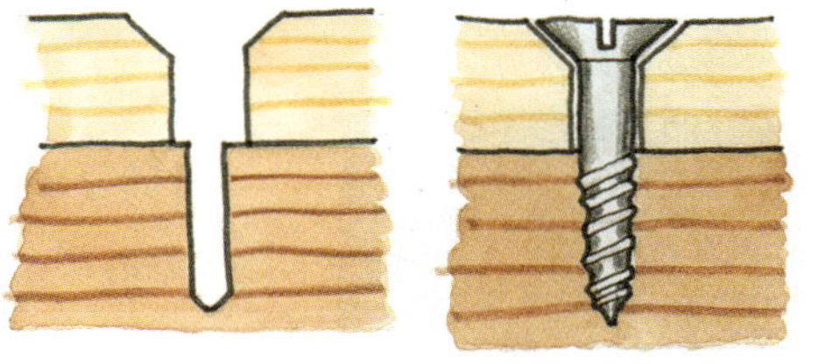

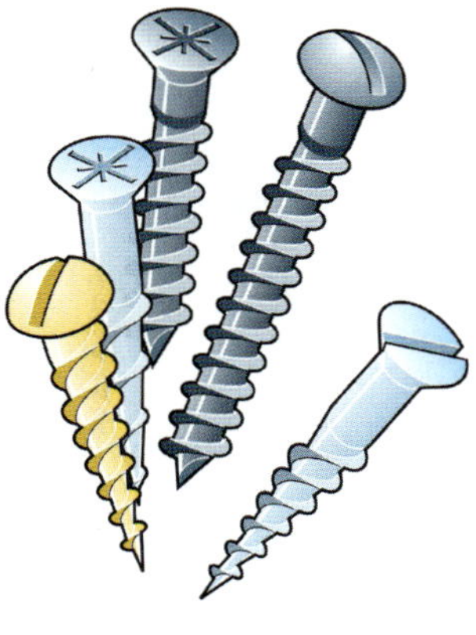

Screw Heads

These come in a variety of types. Decorative press covers are available for **Phillips** and **Pozidrive** screws. **Dome head** screws have a metal dome and are usually used for mirror fixing. **Countersunk headed** screws need a countersunk hole for a flush surface finish.

Written Question

Spend about 8 minutes answering the following question. You will need some paper and something to write with.

Using words and drawings, explain fully the four main steps for joining two pieces of wood together using a countersunk screw. *(8 marks)*

Fastening Metal and Plastic

Self tapping screws are used to join thin sheet material together, or join to thin sheet metal or plastic. There are two types:

- **Thread forming**, which are pointed and are used for soft materials.
- **Thread cutting**, which are blunt-ended and can be used on all materials.

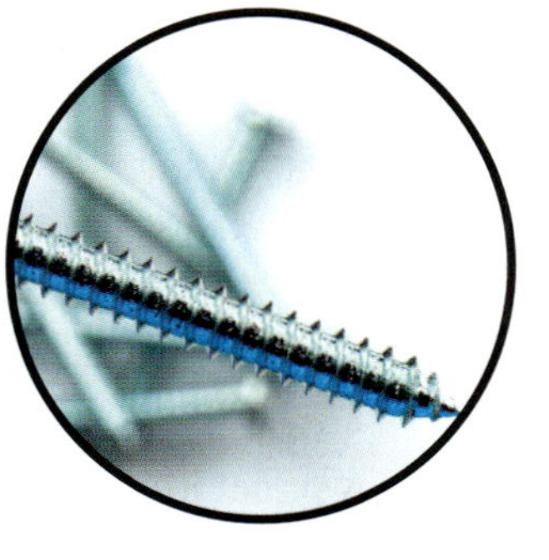

Here's what you need to know...

about the use of nails, pins and rivets.

See *Design and Make It! Resistant Materials Technology* Revised pages 62, 100 (54, 92 earlier edition).

KEYWORDS

Do you know what the following terms mean?

- Nails and pins
- Rivets

Nails and Pins

Nails are metal pins with heads. They are used to join two pieces of wood and/or manufactured board together. The head stops the nail from being pulled through the top piece of wood or manufactured board if the two pieces are pulled apart.

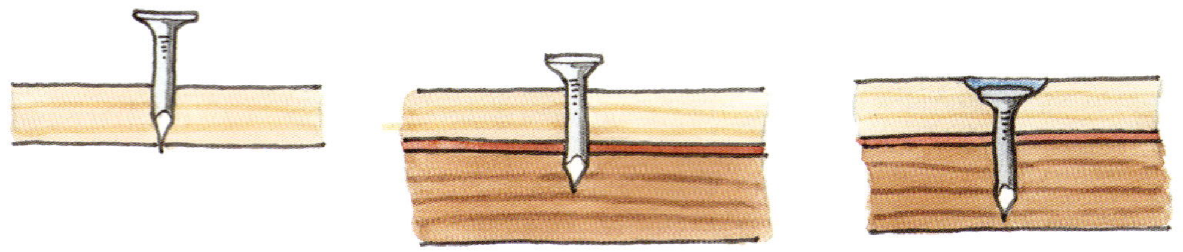

There are many different types of nail that can be specified, but the most common are:

- **Round wire nail**: used for general carpentry.
- **Lost head wire nail**: used for general carpentry. The head is not exposed, so should be used where appearance matters.
- **Oval wire nail**: oval, so less likely to split wood.
- **Panel pin**: general purpose. The head can be punched beneath surface of wood and filled.
- **Hardboard nail**: used for hardboard. Made of copper, it will not discolour paint.
- **Staple, square**: used for upholstery.
- **Tack**: used for upholstery.
- **Decorative upholstery nail**: used for exposed upholstery fixings.
- **Gimp pin**: used for fixing upholstery braid.

In the building industry there are special nails for almost every job, from attaching laths in a roof (**lath nails**), to rough fixing into masonry (**cut clasp nails**).

Rivets

Rivets are a quick and convenient method that is often used to join sheet metal together. They are usually made from aluminium, but also come in copper and stainless steel.

Common diameters are 3, 4 and 5mm and they are available in three different lengths. They are often used with backing **washers**.

To fix rivets a hole needs to drilled that is just big enough to allow the rivet to pass snugly through it. The rivet is then inserted using a **pop riveter** or rivet pliers.

Written Question

Spend about 6 minutes answering the following question. You will need some paper and something to write with.

Different nails and pins are designed for different purposes. Explain what each of the following types are intended to do. Give an example for each answer.

i) A lost-head wire nail. *(2 marks)*
ii) A decorative upholstery nail. *(2 marks)*
iii) A panel pin. *(2 marks)*

Here's what you need to know...

about fixings and fittings.

See *Design and Make It! Resistant Materials Technology* Revised page 120 (112 earlier edition).

KEYWORDS

Do you know what the following terms mean?

- Fixings and fittings
- KD (knock-down fittings)
- Mechanical fastenings
- Hinges

WWW.

Go to: www.artex-rawlplug.co.uk/immediacy-663

Fixings and Fittings

There is a vast range of ready-made **fixings** and **fittings** available in DIY shops. Sometimes a manufacturer will design specific fittings for use with their own products. These fittings save time and money for manufacturers who do not have to invest in people and machinery to make traditional joints.

Knock-down Fittings

Modern furniture is often made using manufactured boards and is available as self-assembly packs. This method of producing furniture has led to a revolution in **KD** or **knock-down fittings**. These are designed to be as 'user-friendly' as possible.

KD fittings also mean that furniture can be produced in **flat packs**. In this way, storage space at shops and factories is saved, and transport costs from factory to showroom are reduced. Many items can be taken home from the shops by the customer, saving the cost of delivery.

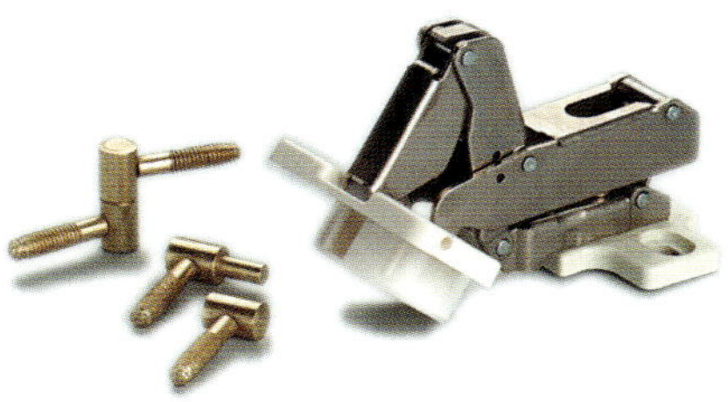

Fixings and Fittings for Wood

Examples of KD fittings for wood are holes which are pre-drilled for screw fittings, plates and inserts already attached to the individual pieces, and multiple holes drilled for a choice of door and shelf positions. Many of these types of construction are held together using dowel pegs.

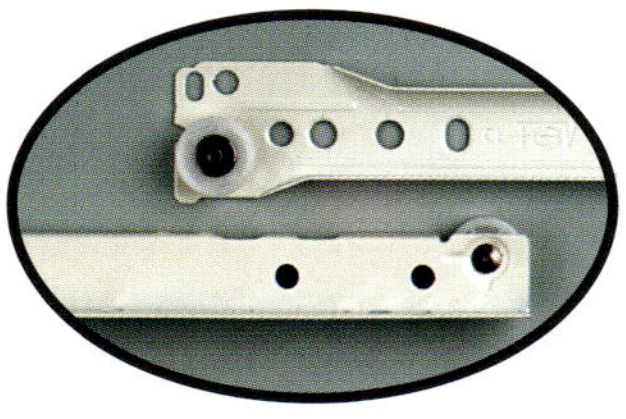

Fixings and Fittings for Metal

There is a wide choice of **mechanical fastenings** for metal and the method chosen will depend on a number of factors. The physical conditions in which the product will be used, for example, the levels of humidity and the physical characteristics needed by the joint, for example, corrosion, are important.

Hinges

Hinges come in a variety of types for different uses. The main types are butt, piano, back flap, and flush. Rising butt and tee hinges are used on doors.

Written Question

Spend about 8 minutes answering the following question. You will need some paper and something to write with.

Modern flat-pack furniture that uses KD (knock-down) fittings offers advantages to both consumers and manufacturers of the furniture.

i) Give two advantages of flat-pack furniture using KD fittings for the consumer. *(4 marks)*

ii) Give two advantages of flat-pack furniture using KD fittings for the manufacturer. *(4 marks)*

Here's what you need to know...

about adhesives.

See *Design and Make It! Resistant Materials Technology* Revised pages 63 and 102 (55 and 94 earlier edition).

KEYWORDS

Do you know what the following terms mean?

- Natural adhesives
- Synthetic adhesives
- Polymers

WWW.

Go to:
www.azom.com/Applications.asp

Adhesives

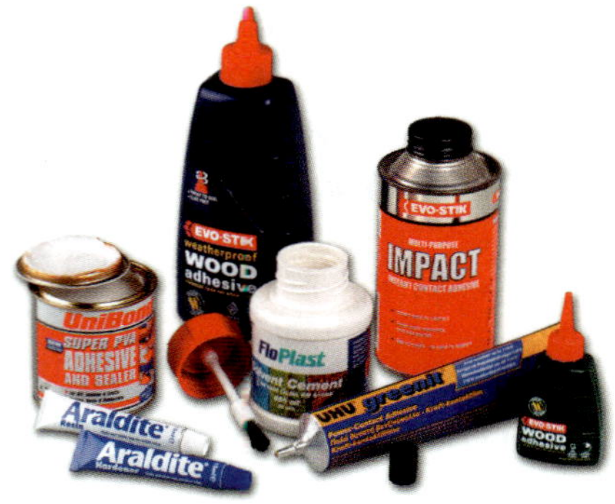

Adhesives hold two surfaces together in what is usually a permanent joint. As the range of manufactured materials is developed, so too has the need for a range of specialised adhesives to join them with. Designers need to make sure they specify the appropriate adhesive.

There are a wide variety of adhesives, both natural and synthetic.

Natural adhesives (glues) are animal or vegetable-based products, made, for example, from bones, fish and plant extracts.

Synthetic adhesives are mainly **polymers**. They have strength and flexibility. Thermoplastic resin-based adhesives are softened by heat.

- **Epoxy resin** (Araldite) is very adaptable. It dries within 48 hours. It is is best for joining metal to wood, metal to metal and acrylic to wood, metal and acrylic.
- **Urea formaldehyde** is a two part adhesive.
- **Synthetic resin**, such as Cascamite is a dry powder that is mixed with cold water.
- **PVA** (polyvinyl acetate) (Resin W) is white, ready mixed liquid. It is best for joining wood to wood and for joining expanded polystyrene anything!
- **Contact adhesive** (Thixofix) is a synthetic rubber colution that sticks on contact. It is good for joining a wide range of materials: fabric to wood, metal, acrylic and melamine, melamine to melamine, and to wood, metal and acrylic and rubber to rubber and to wood, metal, acrylic, melamine and fabric.
- **Acrylic cement** (Tensol), used for plastics, has strong fumes.
- **Latex fabric glue** (Copydex) is rubbery and white solution that provides an immediate bond and is best for joining fabric to fabric.
- **Hot glue** (Bostick) is an adhesive that bonds as it cools.

Applying Adhesives

There are some important things to remember when applying adhesives. These need to be planned for in the production process. For example:

- surfaces to be joined will need to be clean, dust free and dry.
- equal, or varying parts of resin, hardener or curing agent will need to be mixed together in the correct amount using a clean container or piece of card and a spatula.
- even layers need to be applied to the surfaces to be joined.
- excess adhesive needs to be wiped off immediately.
- the structure needs to be held securely while drying, using a clamp, or masking tape if it is an awkward shape that needs to be supported.

A glued joint relies on two things, a good fit and a clean surface.

Written Question

Spend about 5 minutes answering the following question. You will need some paper and something to write with.

Explain carefully the steps for using epoxy resin adhesive. *(5 marks)*

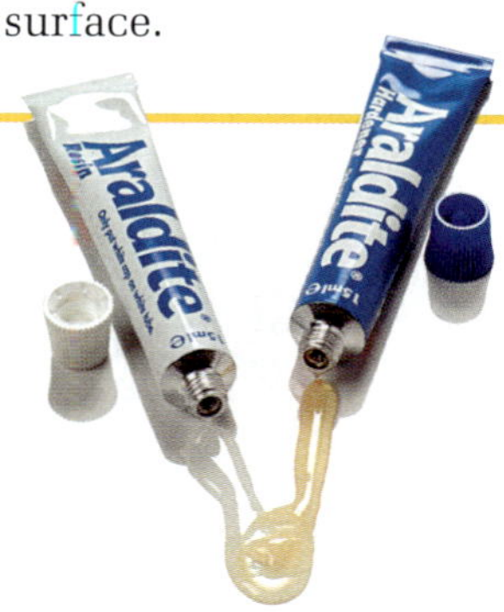

Here's what you need to know...

about the analysis and evaluation of existing products.

See *Design & Make It! Resistant Materials Technology* Revised pages 30–31, 76–77 and 112–115 (22–23, 68–69 and 104–107 earlier edition).

KEYWORDS

Do you know what the following terms mean?

- Product analysis
- Product evaluation
- Design principles
- Critical judgements

www.

Go to:

www.howstuffworks.com/

www.baddesigns.com/canopen.html

The Analysis and Evaluation of Existing Products

Designers study existing products to learn more about how they were designed and made, and to see if they can find ways of improving them. This is known as product analysis and product evaluation.

Product analysis involves studying and describing an object under different headings, such as its size, material or colour, or its different components.

Product evaluation involves testing a product in terms of its effectiveness, e.g. how well it performs, where it might be improved, its strengths and weaknesses, etc. This is known as **performance testing**.

Design Principles

When analysing and evaluating a product, designers study a range of **design principles**. These include its:

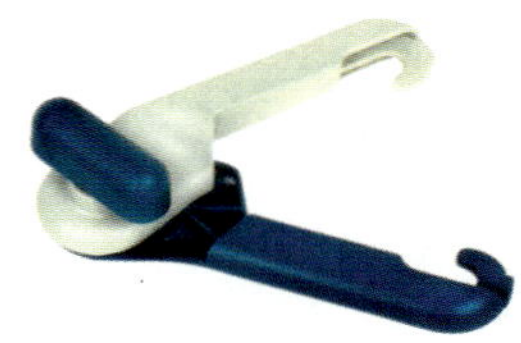

- **form** – its three dimensional shape.
- **function** – what it is supposed to do.
- **materials** – what it is made from and how it is manufactured.
- **component parts** – the different parts it is made up from.
- **colour and texture** – what it looks and feels like.
- **decoration** – added pattern or other parts that are not essential to its working.
- **aesthetic appeal** – the things that make it pleasing to look at, own and use.
- **ergonomic and anthropometric data** – its size and method of operation in relation to the capabilities of its intended users.

Finding Out How Well Products Work

To evaluate how well products work designers devise a number of performance tests. For example:

- how much dust will a vacuum cleaner pick up?
- how easy is it to program a video recorder?
- how durable is a child's toy?
- how comfortable is a chair to sit on for a long time?

Some of these tests are a matter of **fact**, for example, the weight of dust collected, the number of hours a battery lasts, etc. Others are a matter of **opinion**, for example, the look of a piece of jewellery or the appeal of an item of furniture.

Written Question

Spend about 4 minutes answering the following question. You will need some paper and something to write with.

You have been asked to analyse and evaluate the design and manufacture of a child's pushchair. Suggest and outline two performance tests you could carry out to learn more about its stability and portability.

(4 marks)

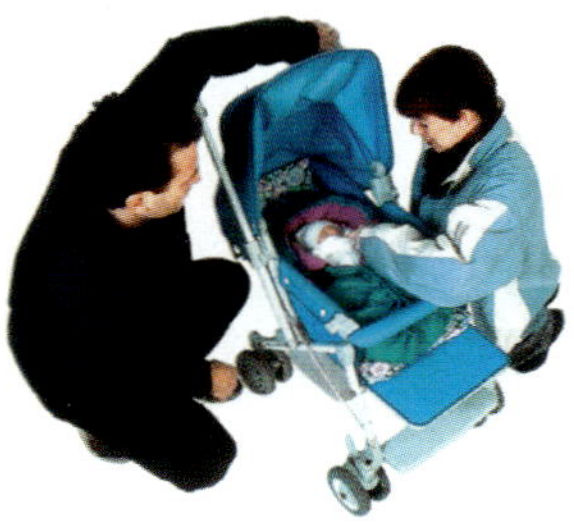

Making Critical Judgements

When all the information about a product has been gathered together a final **judgement** can be made. Being **critical** is not just about saying what is wrong with something. There may be good and bad points about a design – it may work well in one respect, but badly in another.

If several designs of a similar product are being evaluated (e.g. a range of torches) it might be possible to compare them and reach a conclusion about which is the most successful overall for a particular application. However, the main purpose of the exercise is to give the designer ideas for other new or improved products.

Here's what you need to know...

about the analysis of aesthetic and functional design requirements.

See *Design & Make It! Resistant Materials Technology* Revised pages 30, 50–51 and 114–117 (22, 42–43 and 106–109 earlier edition).

KEYWORDS

Do you know what the following terms mean?

- Aesthetics
- Function

www.

Go to:
www.dyson.co.uk/tech/dysoncyclone/default.asp
www.designcouncil.org/betterbydesign

Aesthetics and Function

What are Aesthetics?

Aesthetics are all about how people respond to things through their senses. What we see, hear, taste, touch and smell can be pleasant or distressing experiences. We also respond to harmony and contrast.

Different people like and dislike different shapes, colours, patterns, harmonies and contrasts, etc. This is called **personal taste**. Because of this, different product designs are needed to satisfy the tastes of a range of people.

Our preferences for different aesthetic experiences are partly determined by our **lifestyle** – the particular things we choose to do, where we do them and the products we use. While one person might look for aggressive, angular and brightly coloured shapes and forms that suggest danger and adventure, another person might be more interested in softer, smoother, more subtle shades that suggest comfort and security.

It is important for designers to take potential users' aesthetic preferences into account when developing a new product.

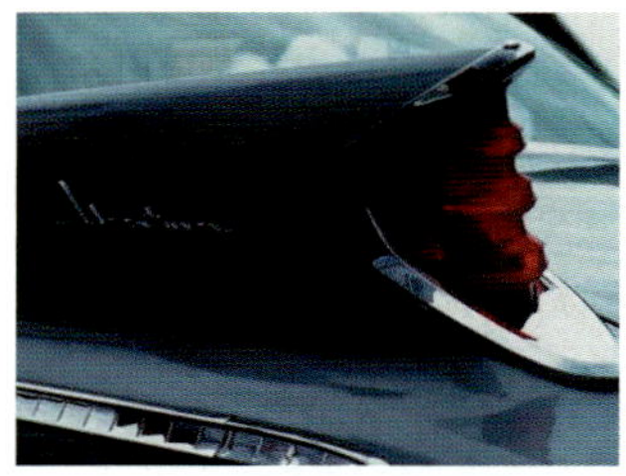

What is Function?

The **function** of a product means the things it is intended to do. A cup is made to hold liquid. A watch needs to tell the time accurately, and be small and light enough to wear on your wrist.

Many objects have more than one function. For example, cups can also be used to measure ingredients. Other cups are used as trophies or as souvenirs, so aesthetic appeal can be a very important part of an object's function.

Designers need to create products that will perform a wide range of functions for different users.

Written Question

Spend about 7 minutes answering the following question. You will need some paper and something to write with.

i) Explain what is meant by the term 'aesthetics'. *(2 marks)*

ii) Study this photo of a plastic watering can. Identify three functional requirements relating to how well it carries water and two requirements concerning its aesthetics. *(5 marks)*

Here's what you need to know...

about analysing the work of past and present designers.

See *Design & Make It! Resistant Materials Technology* Revised pages 6–9, 30–31, 46–49, 76–77, 112–115 and 117 (22–23, 38–41, 68–69, 104–107 and 109, earlier edition).

KEYWORDS

Do you know what the following terms mean?

- Different times
- Different cultures

www.
Go to:
www.designmuseum.org/
www.vam.ac.uk

Products and Designers from the Past and Present

Learning From the Past

As well as looking at recent or current products, designers look back at things created in **different times** and for **different cultures**. Studying what these products were used for, how they worked and how they were made, gives insights into people's lifestyles in the past and the design solutions that were available at the time. Sometimes these can give insights into ways of solving similar problems today.

Living in the Past

Many of today's designs deliberately take on the appearance of products from the past. Many people like to live and work in surroundings that remind them of products and interiors that were fashionable decades, or even centuries, ago. Indeed, some designs even echo the world of the ancient **Egyptians**, **Greeks** or **Romans**. People are often willing to pay extra to own things that make them think of past times.

For example, this piece of modern furniture shown on the right looks like something that might have been made in the 18th century. It those days it would have been hand-made using high-quality materials, so could only have been afforded by the wealthy. Today's **reproductions** are machine-made in quantity, sometimes using composite and synthetic materials.

This table lamp is also modern. It has been designed and made to look like a style of lamp produced by an American company called ***Tiffany*** at the turn of the 20th century, soon after the introduction of electric lighting. It is very decorative and uses organic shapes and motifs based on plants and animals. Designers like **Rennie Mackintosh** and **Antonio Gaudi** used these types of flowing styles in some of their work. This style of work is called **Art Nouveau**.

The radio on the far left was originally mass-produced in the 1930s using one of the first plastics widely available. Its shapes, forms and colours are very warm and friendly, and many people prefer this to the more angular black and silver 'boxes' that modern electronic products often come in.

Written Question

Spend about 12 minutes answering the following question. You will need some plain paper and something to write and draw with.

i) Identify a product that was either designed and made in a previous time, or is modern, but based on designs from a previous time. It might be something you have at home or have seen in a book or magazine. Produce a 2D or 3D coloured sketch of it. If necessary add notes to help describe it. *(8 marks)*

ii) State from what past period you think it has been influenced by, and explain what colours, forms, etc., identify it as coming from that particular era. *(4 marks)*

From Past to Present

Other products we use today are not direct copies of products from the past, but are freely **influenced** by them. Sometimes they mix shapes, colours and materials from products from a number of different eras and create new forms that could only be produced today.

For example, this modern salt and pepper set uses colours and patterns taken from the 1920s, which in turn were influenced by ancient Egyptian designs. They have now been applied to forms that could only be produced in quantity using today's production technologies.

Here's what you need to know...

about using freehand drawings to sketch products in 2D and 3D.

See *Design & Make It! Resistant Materials Technology* Revised pages 19, 94–7 (86-9 earlier edition)

KEYWORDS

Do you know what the following terms mean?

- Annotation
- Plan
- Elevations
- Orthographic projection
- Two-dimensional (2D)
- Three-dimensional (3D)
- Isometric drawing
- Perspective
- Crating
- Scale

WWW.

Go to:
www.bentley.com
(go to the Gallery)
www.adobe.com
www.cadopia.com/

Using Sketches

Designers often use **sketches**, both to draw existing products and to develop ideas for new ones. These sketches can then be shown to, and discussed, with other members of the design team.

Sometimes they find it difficult to keep up – their thoughts flow quicker than they can sketch. There certainly isn't time to use rulers, technical drawing aids or a computer!

Adding Labels

Another way to speed up the process of recording a product on paper is to use **annotations**, or 'labels', alongside a sketch. For example, it might be quicker to label part of a product sketch with the words 'dark green', than it is to find the right shade of pencil or felt-pen to colour it in. As well as words, **numbers** can be added to indicate quantities, sizes, etc.

2D Freehand Drawing Techniques

Designers use a variety of methods to show the product they are sketching. They might sketch a **plan** or an **elevation**, or both together (i.e. a rough **orthographic projection**). These are 2D freehand drawing techniques.

3D Freehand Drawing Techniques

Alternatively, designers might use 3D freehand drawing techniques, using simple **isometric** or **perspective** projections. These are more difficult, so they use the method of '**crating**'. This involves sketching a rectangular box and fitting the product being sketched inside it. Curved shapes and forms can be contained in a 'crate'. Another approach is to use a wire-frame model.

Specially printed grid-paper can be used as an underlay to make construction of 3D sketches easier.

Other Freehand Drawing Techniques

Remember, it is not always necessary to sketch the whole product each time. Designers often focus on one part of the product and just sketch that. They may sketch the part much larger than it will be in reality, i.e. change its **scale**. Other techniques they use include '**cut-away**' sketches and '**exploded**' sketches.

Written Question

Answer the following question on plain paper. Do not spend more than 20 minutes drawing your answer.

You have been asked to sketch some initial design ideas for a new torch. It is to be used as a 'prop' in a science fiction film set in the near future.

Produce at least two annotated sketches that show the development of one idea. Include at least two of the following techniques:

- an orthographic freehand drawing (2D plan and elevations).
- a 3D freehand drawing (isometric or perspective).
- a freehand drawing showing a detail.
- a cut-away or exploded freehand drawing.

As you sketch your ideas you should include notes about:

- how the torch will be held.
- how it will be switched on and off.
- what shape and size of bulb will be used.
- how the source of power can be replenished.
- the materials that might be used, and how it might be made.

Note that marks will be awarded primarily for the appropriate use and quality of your sketching and the techniques you use rather than for the product you design. *(20 marks)*

Here's what you need to know...

about design specifications.

See *Design & Make It! Resistant Materials Technology* Revised pages 9 and 16–17 (10–11 earlier edition).

KEYWORDS

Do you know what the following terms mean?

- Design specification
- Criteria

Design Specification

The **design specification** is a list of requirements or **criteria** for the final product. If these requirements are not clearly expressed, understood and agreed by everyone involved from the start, then the final product is unlikely to be successful. The last thing anyone wants is something that is expensive to make, that no one wants and that doesn't work as intended.

The design specification has a big influence on the design of products. This list of requirements for a product helps ensure that the **client**, **manufacturer** and **user** all get the product they want, i.e. one that is profitable, saleable, easy to make and works well in use.

Requirements

The design specification will cover a wide range of requirements, from things like **size**, **weight**, **appearance** and **function** to **durability**, **safety** and **environmental standards**. The specification states what aspects of the design are **fixed** (e.g. 'must be 30mm wide'), and where there is room for the designer to make decisions within acceptable **limits** (e.g. 'must be between 400 and 500gm in weight').

The client will want to be sure that the specification covers necessary requirements to make sure that the product will be desirable to the particular target consumer group, and can be sold at a cost that will be profitable over a certain time-scale or production run.

The manufacturer will want to be sure that, as far as possible, the materials required are ones that are readily available, and that they can be worked using existing equipment without the need for further training.

The client, manufacturer and designer will all want to be sure that the product is suitable for the potential customer, and works in the way it is intended to.

Checking Design Proposals

During the product development process the designer will come up with many ideas and proposals. These need to be regularly **tested** and **checked** against the design specification to ensure they meet the requirements of the client, manufacturer and user. If for any reason they don't, they need to be redeveloped until they do meet the specification.

The final quality of a product is largely dependent on the quality of the original design specification that was produced.

Design Specification
The mug should be an unusual shape and appeal to children

Written Question

Spend about 55 minutes answering the following question. You will need some plain paper and something to write and draw with.

a) The design team for a new clock are preparing a draft design specification checklist for a meeting with the client and manufacturer. The following have already been decided:

1 The clock must appeal to the 8–14 year age group.
2 10,000 will be manufactured.
3 It must be cheap to produce so it can be sold at a low price.

To achieve this a minimum number of features will be included.

For each of the three statements (1–3) above, explain to what extent it provides the designer with restrictions or opportunities to take decisions about the clock. *(6 marks)*

b) Use the information in the specification above to help you sketch three different design ideas for a clock. Add notes to explain your sketches and provide an evaluation of each idea.

A total of 18 marks will be awarded for:

- three different ideas *(3 x 3 marks)*
- quality of sketching *(3 marks)*
- quality of notes *(3 marks)*
- quality of evaluation *(3 x 1 marks)*

c) Choose one of your ideas from b).
   i) Use notes and sketches to show how the clock would be manufactured. *(15 marks)*
   ii) Use freehand drawing techniques to produce an orthographic sketch of the clock, with dimensions, and a coloured 3D sketch. *(15 marks)*

Here's what you need to know...

about ensuring quality by evaluating product design and manufacture.

See *Design & Make It! Resistant Materials Technology* Revised pages 9 and 20–25 (13–17 earlier edition).

KEYWORDS

Do you know what the following term means?

- Quality assurance

www.

Go to: www.designcouncil.org/betterbydesign

Testing and Evaluation

Quality Assurance

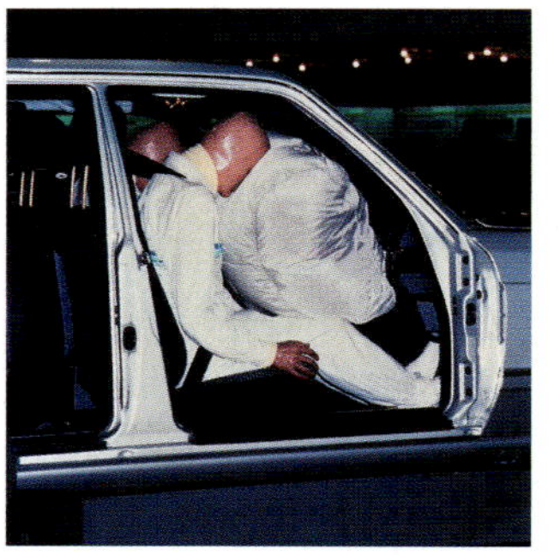

Manufacturers and designers need to ensure that the products they design and make are of good quality. The term '**quality assurance**' is used to describe all the precautions, procedures and tests undertaken before, during and after production. This is discussed further in Topic 8 Section 4.

Testing It Out

During the development of a new product designers are constantly testing and evaluating their ideas to identify and suggest areas for improvement and refinement. They use a variety of techniques and methods to achieve this.

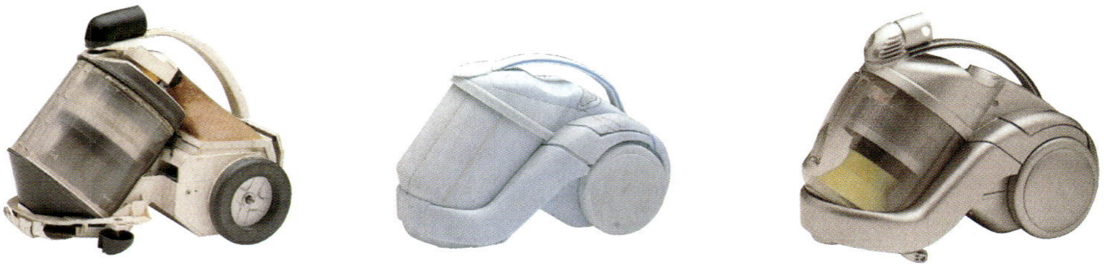

Models, **mock-ups** and **prototypes** of new designs are regularly tested against the requirements of the specification. These tests might be **physical**, e.g. concerned with size, strength, weight, or more a matter of **opinion**, e.g. ease of use, colour, texture, etc. Such tests often need to be specially devised according to the product and the information required.

These models, mock-ups and prototypes may be of a different **size** to the final product, or perhaps made in a different **material**. Sometimes only parts of the product are modelled in order to test out a specific aspect of the design. ICT is often used a great deal in the collection, analysis and presentation of this information.

It's Not For You

During these tests the design team needs to remember that the product is not for their own personal use. They need to consider the needs and views of the client, manufacturer, retailer and the target user. Will the new product be better and more desirable than others already on the market? During the development phase, **surveys** and **interviews** will be undertaken to assess how well a new product compares with others.

Better Make it Better

Methods of production will also need to be tested to find the most effective. Different materials and processes might be quicker, cheaper or more accurate and reliable. This sometimes involves making special tools, jigs and measuring devices.

Written question

Spend about 6 minutes answering the following question. You will need some paper and something to write with.

Study the photograph on the right of a working 3D model of a design for a vacuum cleaner. Suggest three ways in which the working 3D model could be used to test out different aspects of the design. *(6 marks)*

Here's what you need to know...

about social, cultural and moral influences on design.

See *Design & Make It! Resistant Materials Technology* Revised pages 6–7 and 141 (133 earlier edition).

KEYWORDS

Do you know what the following terms mean?

- Social issues
- Cultural issues
- Moral issues

Social, Cultural and Moral Influences

Designers need to take into account the impact the products they create will have on the society and culture in which they are used. There are also moral issues to consider.

Social Issues

In Design and Technology, **social issues** can arise when a new product has an unforeseen side effect on a group of people. This can be a good or a bad thing. For example, the rise in use of mobile phones to send text messages has increased the demand for such phones. This:

- provides employment for factory workers.
- enables people to keep in touch more easily.
- means that there is less need for people to meet each other in person.

Cultural Issues

Cultural issues can arise when a new product does not take into account the fact that a particular shape, colour or name can have very different meanings to different groups of people.

Designers need to take care not to offend groups of people with different traditions and beliefs. For example, different cultures celebrate religious festivals in different ways.

However, a careful choice of name, shape and colour can help promote a sense of unity between different global cultures.

Moral Issues

Should children be stopped from playing with water pistols?

Moral issues occur when a new product could help someone do something that might be considered undesirable or illegal. For example, children enjoy playing with toy guns, but some people think that this encourages them to be more violent. So, designing a toy gun might raise moral issues.

Another example might be designing a bugging device that could invade someone's privacy. This would be considered by the majority of people to be morally bad. However, designing an environmentally friendly product is generally considered to be morally good.

Written question

Spend about 10 minutes answering the following question. You will need some paper and something to write with.

i) Explain what you understand by the terms:

a) a social design issue. *(2 marks)*
b) a cultural design issue. *(2 marks)*
c) a moral design issue. *(2 marks)*

ii) Give an example of a specific existing product design and briefly discuss the social, cultural and/or moral design issues that might have been involved. *(4 marks)*

Here's what you need to know...

about ergonomics.

See *Design & Make It! Resistant Materials Technology* Revised pages 80 and 114 (72 and 106 earlier edition).

KEYWORDS

Do you know what the following term means?

- Ergonomics

www.

Go to:
www.baddesigns.com/examples.html

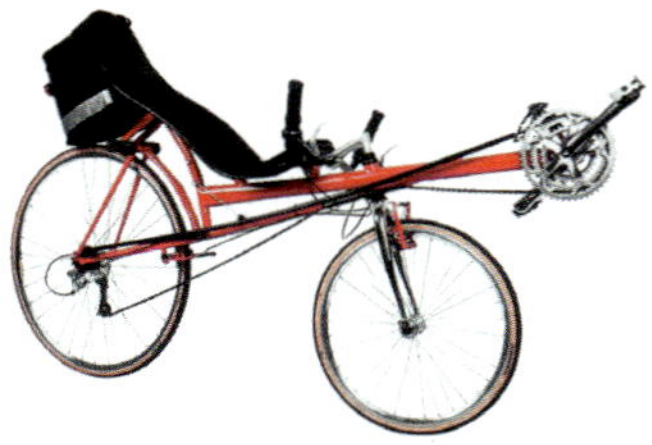

What is Ergonomics?

Ergonomics is the study of people in relation to their living and working environment. Ergonomics helps designers make everyday things that:

- are the best size and shape.
- have their controls and displays in the right place.
- are easy and satisfying to use.

Ergonomics concerns both what people can do **physically** – how easily they can turn a handle, for example – and how their **senses** respond to the appearance, touch, sound, smell and taste of a product.

The Same, But Different

Creating products that worked well would be easier if everyone was the same. Unfortunately, people come in a wide range of **sizes** and **physical capabilities**, and react to different things in different ways.

Some products are aimed at a very specific market, such as a toy for a child under 3 years, which makes things a bit easier. Special products are designed for the elderly or handicapped to meet their particular requirements. Others need to be suitable for a much wider consumer group – a door is a good example of a product that has to meet the requirements of a high percentile of the population.

One solution to the problem is to provide a range of products in different sizes and/or appearances. For example, bicycles can have different frame, handlebar and wheel sizes, and come in a range of colours. In this way they become suitable for a wide range of people, from a fashion-conscious teenager to a middle-aged racing enthusiast, or even for someone with special needs.

Collecting Ergonomic Information

There are two main ways in which a designer can obtain information about the ergonomics involved in a particular product. One is to consult information that has already been documented about people's physical and mental capabilities in undertaking specific tasks. For example, someone designing a vacuum cleaner would need to find out a lot about how users grip the handle in order to steer the machine, what weight they can comfortably carry upstairs, what height to make the handle, etc.

The other method is to consult existing statistical data about ranges of people's physical sizes and capabilities. This is known as **anthropometric data**. This is discussed in more detail in the next section.

If the information they require is not readily available the designer may need to devise a series of tests themselves. This is likely to involve observing and measuring a range of different people using similar existing devices, or specially built mock-ups of new designs.

A great deal of information can be gained from the analysis of **existing products** to identify how other designers have solved similar problems.

Written Question

Spend about 8 minutes answering the following question. You will need some paper and something to write with.

i) What do you understand by the term 'ergonomics'? *(2 marks)*

ii) As a designer you have been asked to undertake an ergonomic evaluation of the telephone shown on the left. Identify three things you would test. *(6 marks)*

Here's what you need to know...

about the use of anthropometric data in design.

See *Design & Make It! Resistant Materials Technology* Revised pages 81 and 141 (73 and 133 earlier edition).

KEYWORDS

Do you know what the following terms mean?

- Anthropometrics
- Physical capabilities

WWW.
Go to:
www.design-technology.info/anthropometrics/default.htm

What is Anthropometrics?

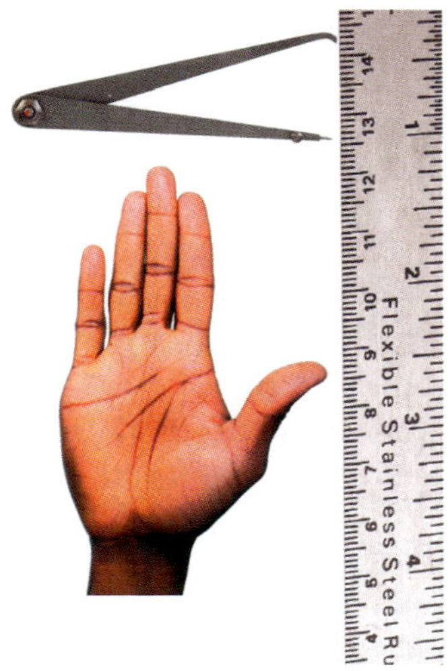

Anthropometrics is the scientific measurement of the human body. It is concerned with gathering statistical data about the dimensions and **physical capabilities** of different groups of people. Designers use the data for guidance, and it saves them having to collect large amounts of measurements for themselves.

The data is not just about the overall height or width of people, but provides detailed analysis of all parts of the body, such as the size of hands and fingers. It also involves things like how far people can reach, the weight they can lift, the pressure they can exert, the extent to which the wrist can rotate, what angles of vision they have, etc.

Anthropometric data can be measured in two ways:

- **static**, i.e. when the body is still, e.g. sitting, standing or lying.
- **dynamic**, i.e. when the body is moving, e.g. twisting, stretching, pressing.

Different, But Similar

Anthropometric data is sub-divided into many different **categories**. For example, British males aged 20–29 years, British females aged 20–29 years, Japanese males aged 50–59 years, and so on. Sizes can vary widely according to age, gender and racial origin. Other factors such as diet, injury or disability can also have an effect.

Designing for Mr Average?

Using anthropometric data is more than just using averages of the population. The number of people who are exactly average is very small. The data is mainly used to identify the extremes of size. Generally designers aim to accommodate 90% of likely users. This means that a product can be used by all but the smallest 5% and the largest 5% of the target market.

Anthropometric data is usually published in table form, showing the measurements for different sexes, ages and races at the 50th percentile (i.e. the 'average'), and the 5th and 95th percentiles.

A Calculated Decision

In some situations a designer might commission a specialist company to collect accurate data for a specific situation not included in standard anthropometric tables.

For example, a design team needed to determine the best height for a push button on a calculator, after it had been pressed. They wanted to reduce the amount of surface scratching caused by long fingernails. Measurements of the right forefinger nail lengths of a wide variety of the population were taken to the nearest 0.5mm.

The results revealed that the push button would need to be 4mm above the surface to avoid all scratching. However, only 5% of people had a nail length of more that 3mm, so 3mm was used as an acceptable height.

Written Question

Spend about 9 minutes answering the following question. You will need some paper and something to write with.

i) What do you understand by the term 'anthropometric data'? *(2 marks)*

ii) What are the three main ways in which anthropometric data is usually classified? *(3 marks)*

iii) You have been asked to design a toy for a 3- to 6-year-old to be sold only in the UK. The brief states that it must include a simple wind-up mechanism. Identify two items of anthropometric data that would be helpful to know when choosing the wind-up mechanism. *(4 marks)*

Here's what you need to know...

about consumer choice and rights.

See *Design & Make It! Resistant Materials Technology* Revised pages 9, 92–93 and 139 (84–85 and 131 earlier edition).

KEYWORDS

Do you know what the following terms mean?

- Market research
- Consumer rights
- Fit for purpose

www.

Go to: www.consumereducation.org.uk/

Consumer Choice and Right

Any Colour You Like!

Different people like and want different things. Exactly what we buy is often determined by our age, social status and cultural background. Not so long ago there was very little choice of products in the shops – one style or colour or brand had to do for everyone. Today, however, manufacturers produce a much wider range of goods so we can choose much more closely the product that does exactly what we want it to, and looks the way that closely reflects our taste and individual **lifestyle**.

Finding Out What's Wanted

To discover exactly what people want, extensive **market research** is carried out. **Surveys** are undertaken to discover consumers' habits, attitudes and aspirations. A common method of obtaining such information is a **questionnaire**, targeted at a particular market that the manufacturer is interested in – there's not much point asking a senior citizen about their favourite computer game, or a teenager about pension plans.

Data from such questionnaires can be recorded on a computer database and analysed in sophisticated ways. Understanding **patterns of purchase** can be very useful – for example, if someone buys a certain type of car, they are also likely to be interested in a particular holiday location.

Designers need to ensure that the features, shapes, sizes, materials, colour, etc. of their products will specifically appeal to the **consumer market** it is aimed at.

Consumer Rights

Manufacturers have a moral responsibility to ensure that the products they sell are **fit for their purpose**, i.e. they do what they claim to do, and continue to do so. They must also be safe to use. A child's toy should not be so fragile that it breaks almost immediately, or contain parts that could be dangerous. There are extensive **safety regulations** and standards that designers and manufacturers must legally comply with, such as those produced by the **British Standards Institute**.

Read the Instructions First!

Providing the consumer with appropriate information about the product is also important, such as what it's made from, what it is or is not intended to do, whether it's suitable for young children, etc. Again, there are a range of British Standards to refer to for guidance. Packaging and labels are used to communicate this information

Satisfaction Guaranteed?

The consumer needs clear instructions about what to do if a product is found to be faulty in some way. Manufacturers have a legal and moral obligation to refund payment, repair or replace a product that does not perform as stated.

Written Question

Spend about 9 minutes answering the following question. You will need some paper and something to write with.

On the right are three portable music devices. They are each aimed at different target consumer markets. For each of the three devices, give an explanation of which market it is aimed at and how it achieves it. *(9 marks)*

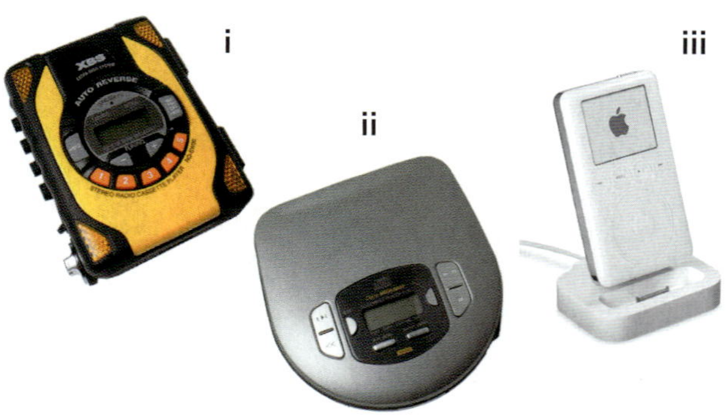

Here's what you need to know...

about environmental design issues.

See *Design & Make It! Resistant Materials Technology* Revised pages 6–7 and 78–79 (70–71 earlier edition).

KEYWORDS

Do you know what the following terms mean?

- Environmental design issues
- Non-renewable energy
- Deforestation
- Reduce, Recycle and Reuse

Green Technology

Making products uses up the earth's resources and can cause **environmental damage**. Waste materials need to be disposed of carefully. Designers, manufacturers and consumers have a moral responsibility to ensure the damage to natural resources is kept to a minimum. **Environmental design issues** are concerned with the impact a new product has on nature and how nature's resources are used.

A product may be made with a high proportion of materials that cannot be replaced, or using chemicals that pollute the atmosphere. This can also be harmful to wildlife and cause changes to weather patterns and the land. Some production processes use high levels of **non-renewable energy**.

Wood

Paper and wood are made from trees. In many parts of the world forests are being destroyed. This is known as **deforestation**. A high percentage of paper and wood can be used again if recycled, reducing the number of trees that need to be cut down. It is also important to ensure new trees are grown to replace the ones cut down. In general terms wood is a very **environmentally friendly material**, as it can be replaced comparatively easily and quickly if managed properly.

Plastics and Metal

Plastics are made from oil and metals are mined from the earth. There are only limited supplies of oil and metal. Converting metal ore into usable material consumes large amounts of energy. The amounts of plastic and metal used in a product need to be reduced to the minimum. Some plastics and most metals can be recycled.

Find a Bin to Put It In

A product may also be difficult to dispose of when it has been finished with. This can cause further environmental problems. An aluminium can takes up to 100 years to **decompose**. A plastic product can take much longer.

The 3 Rs

When designing products it's important to remember the **3 Rs**: **Reduce, Recycle and Reuse**. The 3 Rs provide guidance on how to minimise the damage done to the environment by a product.

Reduce
Use the least amounts of materials and energy in making a product.

Reuse
Use products and components (e.g. containers, electronics, fastenings) that have already been used and/or can be used again in different products.

Recycle
Use recycled materials and/or materials that can be recycled after use (e.g. untreated papers, glass). Recycled materials are those which can be used again in new products. This usually involves separating the materials into different types and then cleaning and re-preparing them.

Written Question

Spend about 6 minutes answering the following question. You will need some paper and something to write with.

On the right is a photograph of a torch. Explain how a designer might have applied the 3 Rs (Reduce, Recycle and Reuse) during the development and manufacture of this product. *(6 marks)*

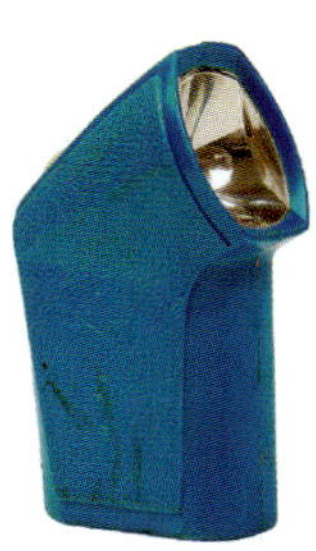

Disadvantages

While these approaches help save the environment, they can bring disadvantages in terms of increased costs of materials, manufacturing and recycling processes, and may result in products being produced that may not perform as well as required. However, the most successful designs manage to combine environmental friendliness with low-cost, quality production.

Here's what you need to know...

about identifying hazards and reducing risks.

See *Design & Make It! Resistant Materials Technology* Revised page 141 (133 earlier edition).

KEYWORDS

Do you know what the following terms mean?
- Hazards
- Risks
- Risk assessment

Safety First

Products need to be safe for the consumer to use (this is discussed further in section 4 of this topic). However, products also need to be safe for someone to make. Potential dangers, or **hazards**, in the factory or workshop include:
- unguarded machinery and tools.
- poor layout.
- lack of training.
- inadequate safety devices, procedures and signs.

Taking the Risk

Accidents in manufacturing can have serious consequences. Unsafe acts or conditions are called hazards.

There is an important difference between a hazard and a risk. A **risk** tells you how likely it is that a problem will happen. A hazard is the problem itself. If something is risky, it means there's very likely to be a hazard involved.

Risk assessment helps to make manufacturing safe by minimising the risk of hazards. Risk assessment is about identifying what hazards might happen, how likely they are to happen, and what the result would be if they did happen. It works out where the risks of hazards are in a process, and how to make things as safe as possible. Different processes and areas of the workplace may be identified as being high, medium or low risk.

Making It Safely

There are a wide range of **regulations** and **guidelines** that manufacturers must follow when asking employees to work with various tools, machines and materials, or when undertaking various processes. As well as guarding against the risk of injury, there are also requirements for the workplace, e.g., lighting, noise, heating, ventilation. All workplaces have a health and safety representative and are visited regularly by a **Health and Safety Inspector** who checks that the necessary **precautions** are being taken.

Design at Work

When developing a production plan for a product, the designer needs to take the requirements for health and safety into account. The use of a particular piece of equipment, material or process might need special monitoring and checking, training and warnings. This might influence the choice of materials and processes specified.

Written Question

Spend about 4 minutes answering the following question. You will need some paper and something to write with.

The wooden toy shown below is to be manufactured in batches in a small workshop. Suggest two safety hazards that could occur during production, and a way of reducing the risk of each happening. *(4 marks)*

Safety in School

Safety procedures and risk assessment are important in school workshops, studios and classrooms too. You need to comply with school **safety rules**, which include legal rules on health and safety. You also need to apply your knowledge of safety to your individual working situation to ensure your own safety and the safety of others. For example, you should recognise:
- the dangers of bladed tools and machines (saws, chisels, drills, etc.).
- the need to open windows/ventilation when 'sanding'.

Using Tools and Equipment Safely

Here's what you need to know...

about using workshop tools and equipment safely.

See *Design & Make It! Resistant Materials Technology* Revised page 141 (133 earlier edition).

KEYWORDS

Do you know what the following terms mean?

- Safety precautions
- Hazards

WWW.
Go to:
www.herts.ac.uk/lis/ltdu/projects/mm2

Workshops and studios can be dangerous places. There are lots of tools and equipment that can cause a great deal of harm if **safety precautions** are not followed.

It's important to think about the tools and equipment you use and the space you use them in. This section focuses on controlling tools and equipment safely. Section 3 of this topic looks at the spaces they are used in.

Using Power Tools

A designer's workshop will probably include a variety of electrically powered drills, small lathes and hand-held tools. All these are potentially dangerous if mishandled. Always stop to think about the specific **safety warnings** your teacher has given you for each one. Also remember the general **workshop rules** that need to be observed.

Using Hand Tools

Even if you are only using hand tools there are many possible **hazards** to be aware of. Saws and chisels have very sharp edges that can easily cut – and skin is less resistant than wood, metal or plastic! Adhesives, spray paints, craft-knives and even scissors can cause injury if not used with **caution**.

Using ICT

Remember too that there are safety considerations involved when using ICT. Never open the computer casing – there are high voltages inside. If you use computers for long periods you may start to suffer from backache, eye strain or aching arms.

Written Question

Spend about 10 minutes answering the following question. You will need some paper and something to write with.

i) On the left is a photograph of a desk-top drill and of a craft knife. Identify two important safety requirements to remember when using each. *(4 marks)*

ii) Briefly describe the potential health risks involved in working at a computer screen for long periods of time. Give three possible symptoms and most likely cause of each. *(6 marks)*

Here's what you need to know...

about creating a safe working environment.

See *Design & Make It! Resistant Materials Technology* Revised page 141 (133 earlier edition).

KEYWORDS

Do you know what the following terms mean?

- Working environment
- Air quality

www.

Go to: www.herts.ac.uk/lis/ltdu/projects/mm2

The Working Environment

Workshops can be dangerous places. There are lots of tools and equipment that can cause a great deal of harm if **safety precautions** are not followed.

It's important to think about the tools and equipment you use and the space you use them in. Section 2 of this topic focused on the tools and equipment. This section looks at the spaces they are used in: the **working environment**.

Placing Tools and Equipment

Careful planning of the layout of workshops and studios can help reduce the risks of **injury**. The positioning of some tools and equipment close together can be **dangerous**.

- For example, placing a computer next to a sink would not be a wise idea as splashing water might lead to electrocution.
- The placing of power points and trailing cables also needs close attention.
- Storage systems need careful thought. Is it easy to reach dangerous items? Are any large piles of a material likely to fall over?

Maintaining Air Quality

Facilities for **ventilation** and fume and dust extraction are also essential considerations.

- Dangerous fumes can be given off when plastic-based products are cut, or when two chemical components are mixed.
- Adhesives, spray paints and varnishes can be dangerous if inhaled.
- Fine dust from cut materials can also be a health problem if not extracted.

Looking After Yourself

Protective clothing, such as goggles, ear protectors and overalls, needs to be readily available. There also needs to be facilities for regular **washing**.

Written Question

Spend about 4 minutes answering the following question. You will need some paper and something to write with.

i) Describe two steps you took to control the risks involved at different stages of the making of your final coursework project. *(2 marks)*

ii) Identify two important safety considerations when planning the layout of a small workshop. *(2 marks)*

Here's what you need to know...

about ensuring that end products are safe for consumer use.

See *Design & Make It! Resistant Materials Technology* Revised pages 92–93 and 139 (84–85 and 131 earlier edition).

KEYWORDS

Do you know what the following terms mean?

- British Standards Institute (BSI)
- Safety mark/symbol

www.
Go to:
www.bsi-global.com/Education/index.xalter

Making It Safe to Use

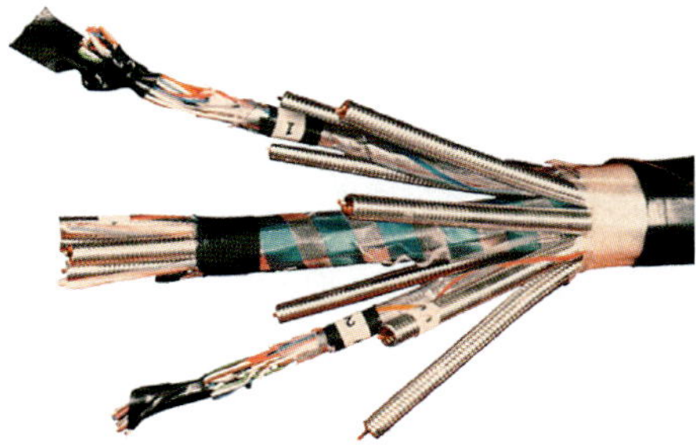

As well as being safe to make, the manufacturer and designer need to ensure that a product is **safe to use**. All products and components are potentially dangerous in some way. For example, the possibility of sharp edges, small splinters, toxic finishes, exposed electrical parts, etc., all need to be eliminated in the design and manufacturing prototype stages. The general stability of a product is important too to ensure it does not fall over.

Maintaining the Standard

The **BSI (British Standards Institute)** provides detailed safety guidance and regulations for a wide variety of products. Their documents highlight the essential technical requirements for a product. If a product is to be sold outside the UK it will also need to conform to the required standards of each of the countries it is to be sold in. Similarly, products made outside the UK must conform to BSI standards if they are to be sold here.

Toy Story

Safety is of particular importance in the design of children's toys. The requirements are covered in BS 5665. The standard has three main sections:

- **physical** and **mechanical** – all the parts that can be touched.
- **flammability** – how quickly it could catch fire
- **choking hazards** – parts that could be detached and swallowed.

A careful choice of materials and finishes is needed, and product stability is also very important.

Testing Times

Before a product can be put on sale it has to be thoroughly tested to ensure it meets the requirements of the BSI standard. The test methods are laid down in the BSI specification, and are often carried out by independent organisations. Sometimes, special **test-rigs** are set up to quickly simulate use over a long period of time, so that the effects of wear and tear can be predicted.

After a product has passed its safety tests the manufacturer can include the appropriate **safety mark** or **symbol** on the product and/or its packaging.

Consumer groups, such as the Consumers' Association, conduct their own tests on a wide range of products, including safety. As well as scientific performance testing, they also interview users.

Written Question

Spend about 7 minutes answering the following question. You will need some paper and something to write with.

i) What do the initials BSI stand for? *(1 mark)*

ii) On the right is a photo of a table lamp. Identify and explain three different potential safety problems that would need to be considered during its design. *(6 marks)*

Here's what you need to know...

about holding devices that are used to secure resistant materials while they are being cut, shaped and joined.

KEYWORDS

Do you know what the following terms mean?

- Woodworking vice
- Engineer's vice
- G clamp
- Sash and mitre clamps
- Mitre box
- Pliers

Holding Devices

Woodworking Vices

Woodworking vices are available in a range of sizes. They are usually permanently attached to woodworking benches, but clamp-on versions are also available. The metal faces should be lined with 10mm ply to protect work that is clamped inside them.

Engineer's Vice (Metal Working Vice)

This is a very strong, heavy vice used for holding metal. The **jaw** openings range from 63mm to 150mm. The jaws often have a rough, non-slip surface and should be used with 'soft jaws' when holding soft metals such as aluminium to prevent marking.

Workmate Bench

This piece of equipment is very useful because it has a full length vice along the top of the bench. It is often used when irregular shaped work pieces need to be secured. It folds flat for storage.

Bench Hook

This slides on to the edge of a woodworking bench and is used to steady short pieces of timber against the back rail when you are sawing them.

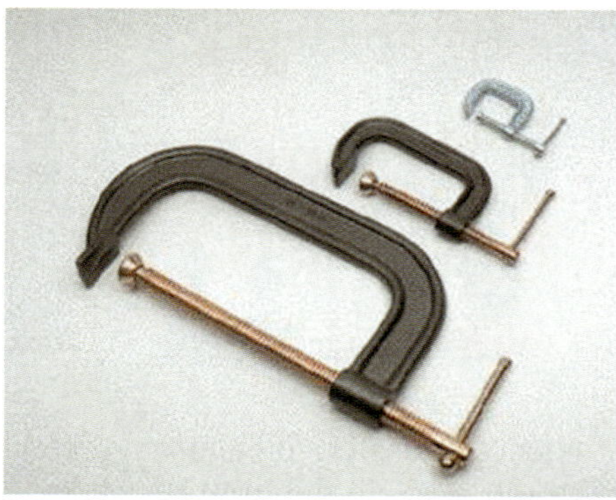

G Clamp

These are used to hold pieces of wood or metal that are too large for a vice or to hold down pieces for drilling or shaping. They range from 50mm to 300mm openings.

Sash and Mitre Clamps

These are specialist tools that are used to hold lengths of wood at the correct **angle** while they are being glued.

Mitre Box

The slots in the box guide the saw to cut through wood at the required angle. They come in a range of sizes and materials.

Pliers

Pliers come in a range of sizes and jaw shapes. They are used to grip and bend wood, metal and plastics.

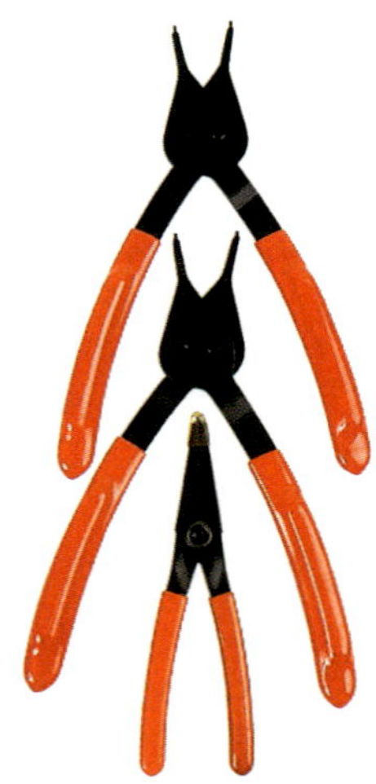

Written Question

Spend about 8 minutes answering the following question. You will need some paper and something to write with.

You are making a door number. Describe in detail, using notes and a sketch drawing, how you would hold a piece of aluminium 30mm by 50mm while you were sawing a slot in it. *(8 marks)*

Here's what you need to know...

about marking out tools.

KEYWORDS

Do you know what the following terms mean?

- Try square
- Mitre square
- Scriber
- Sliding bevel
- Combination square
- Marking gauge
- Mortice gauge
- Face side
- Face edge

Marking Out Time!

Accurate **marking out** means materials can be cut to the right size and shape. There are various techniques, using a variety of tools, for different materials.

Try Square

This is used with a pencil (never use a pen as it will stain the wood grain) to mark out accurate right angles (90° angle) on wood and to check angles in frames. The **mitre square**, used to mark out mitre joints, has a blade at a 45° angle.

Engineer's Try Square

This is used with a **scriber**, which is a thin, pointed metal rod, for marking out metal and plastic. Marking blue is sometimes applied to the surface of metal before scribing to make the mark show up more clearly.

Sliding Bevel

This has an adjustable blade and is useful for marking repeat lines at the same angle on all materials.

Combination Square

This is a steel rule clamped on to a square head. The head can be locked at any point on the rule. The head can be used as both a try square and a mitre square.

Marking Gauge

This is used to mark out on wood, metal and plastic. A wooden stock can be fixed at any point on a beam with a **thumbscrew**. A steel pin projects from the base of the stock and this is used to mark the work piece.

Mortice Gauge

Like the marking gauge but with two pins. It is used to mark out mortice and tenon joints on wood.

Measuring Tools

The **steel rule**, **safety rule** and expanding **metal tape measure** are all accurate ways of measuring your work. Avoid using wooden and plastic rules as these can be easily damaged and can change shape or length slightly in damp or hot conditions.

Written Question

Spend about 8 minutes answering the following question. You will need some paper and something to write with.

Describe in detail, using notes and sketches, the stages of marking a line around a rectangular piece of wood needed for a folding wooden stool. Make sure you name each tool used.

(8 marks)

Marking Knife

An accurate way to mark wood across the grain. It is used instead of a pencil.

Marking a line on a piece of wood

Make sure the **face side** (the wider side) and the **face edge** (the narrower side) have been clearly marked. The stock of a try square is held against the face edge and a line is marked with a pencil after its distance from the end is measured with a steel rule. A line is then marked on the face side, end-on to the last line. Two more lines are drawn on the other edge and face. All four lines should join up.

Here's what you need to know...

about cutting tools.

KEYWORDS

Do you know what the following terms mean?

- Sheet saw
- Back saw
- Frame saw

Saw Points

Saws are used for cutting materials to size. They are used for wood, metal and plastics. There are three main types of **handsaws**:

- **sheet saws** – usually larger handsaws which have a deep but thin blade: for rapid cutting.
- **back saws** – have a folded piece of metal along the top of the blade to stop it from bending: for accurate cutting.
- **frame saws** – usually have very thin blades most of which can cut round corners (coping saw, fret saw). Hacksaws are used for cutting metal and do not cut round corners. The frame stiffens the blade by keeping it in tension.

Hand Saws

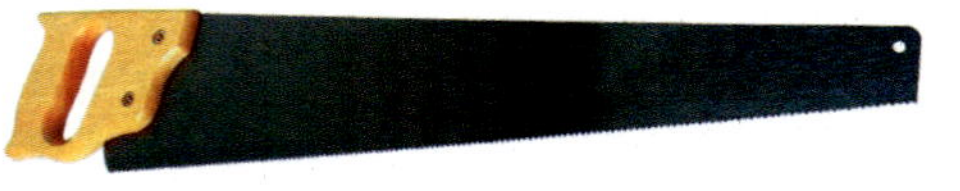

These saws are sheet saws and are used to cut wood. They come in three different types: the **rip saw** is used to saw along the grain, the **crosscut saw** is used to cut across the grain and the panel saw is used to cut manufactured boards. The difference between them is the shape, size and position of the teeth.

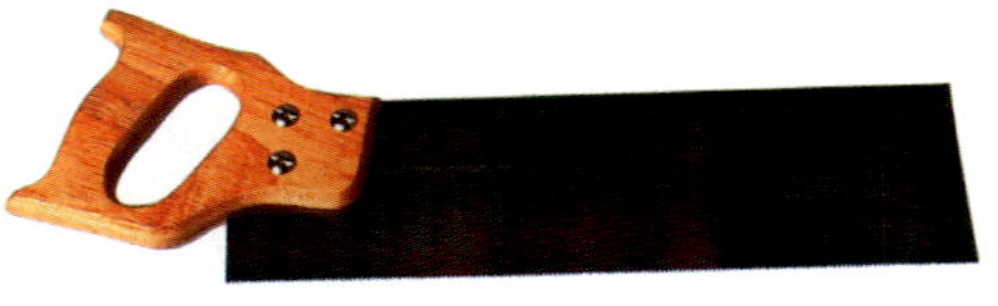

Tenon Saw

This is an example of a back saw. The blade has a folded piece of metal along the top of the blade to stop it from bending. It is used to cut accurate joints in wood.

Dovetail Saw

This is like a tenon saw, but is smaller in size, has finer teeth and is used to cut complicated or very fine joints.

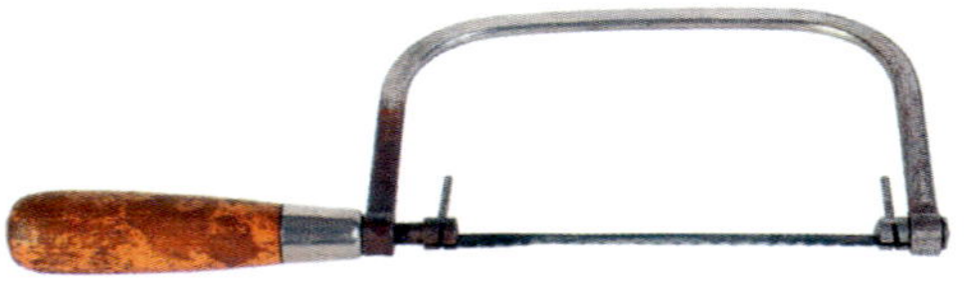

Coping Saw

This is a frame saw. It is used for cutting wood, manufactured boards and plastics. The frame holds the blade in tension, but the blades are fragile and snap quite easily. The frame and slotted pins allow the blade to be turned to saw in any direction.

Fret Saw

This saw has a deep frame and fine narrow blade. It is used to cut complicated shapes in sheets of wood and manufactured board.

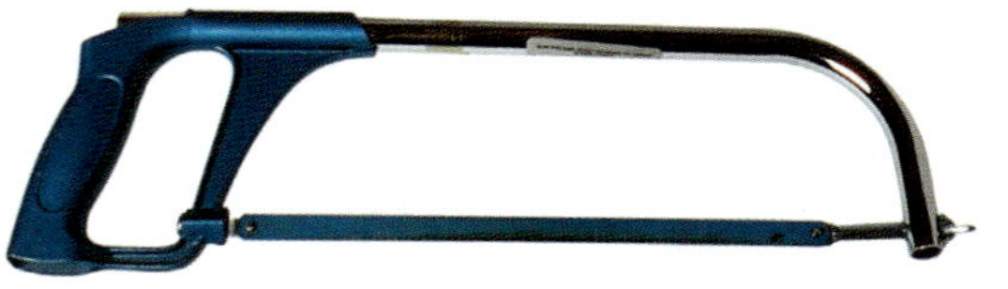

Hacksaw

This is an example of a frame saw that is used to cut metal. They come in two sizes, 255mm and 305mm, but most will take blades of both sizes.

There are four different grades of blade, based on the number of teeth per 25mm. The tension in the blade can be adjusted.

Junior Hacksaw

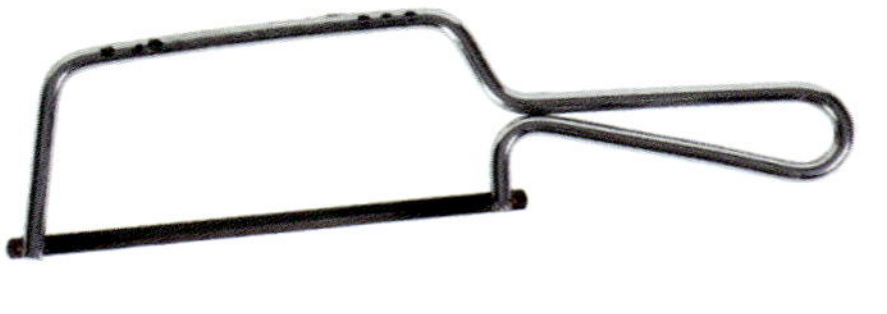

This is a frame saw. The blades come in one size: 150mm. It is used in the school workshop as a general purpose saw for wood, metal and plastics, but was developed for use with metal.

Written Question

Spend about 8 minutes answering the following question. You will need some paper and something to write with.

Describe in detail, using notes and sketches, the stages for using a coping saw to cut curves in sheet material safely.
(8 marks)

Here's what you need to know...

about drilling tools.

KEYWORDS

Do you know what the following terms mean?

- Torque
- Chuck
- Drill bit
- Tang

Drilling Tools

Hand Drill

This is used for drilling wood, metal and plastic. It has a large gear wheel and one or more pinions which are used to apply **torque** (twisting effect) to the **chuck**, which holds the **drill bit**.

Bits are fitted by holding the large gear wheel, turning the chuck anticlockwise to open the jaws, inserting the bit and tightening the jaws around the bit and by turning the chuck clockwise.

Carpenter's Brace

This is used with **auger bits** to cut large holes in wood and manufactured boards.

- Sometimes there is a ratchet behind the chuck. This allows for a part turn when space is limited.
- Bits have a square shaped **tang**, which is gripped by a 'v' groove in the jaws.

Drill Bits

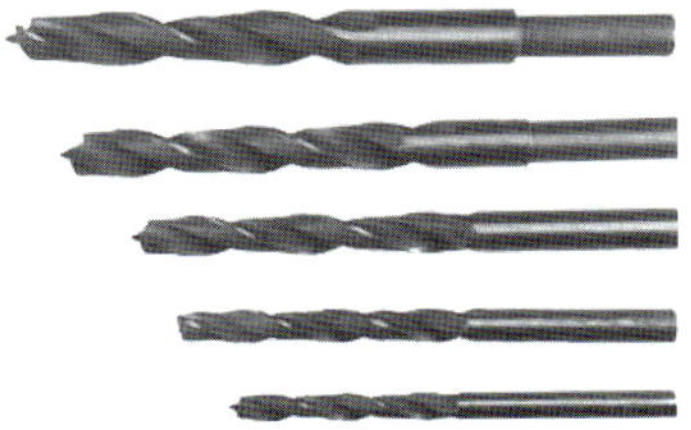

There are a range of different types and sizes and diameters of drill bits. The correct one needs to be chosen according to the size and depth of hole, the material and the drill being used.

Drilling Holes

The position of the hole is marked out with a pencil on wood or scriber on metal. The work piece is secured with a G clamp and the workbench protected with scrap wood.

The hole is started with a **bradawl** for wood, or for metal the hole is **centre punched**. Large holes need to be started with a smaller **pilot hole**.

The drilling needs to be done slowly and carefully, letting the drill bit do the work. When finished, the direction of drilling is reversed, and the drill gently lifted up out of the hole. When drilling metal, a lubricant may be needed.

Written Question

Spend about 8 minutes answering the following question. You will need some paper and something to write with.

Describe in detail, using notes and sketches, the stages for safely drilling a hole in a piece of wood by hand during the manufacture of a pencil rack. *(8 marks)*

Here's what you need to know...

about files and abrasive papers.

See *Design and Make It! Resistant Materials Technology* Revised page 104 (96 earlier edition).

KEYWORDS

Do you know what the following terms mean?

- Files
- Abrasive papers
- Coarseness
- Cut
- Safe edge

Files and Abrasive Papers

Smooth Operators

Files are hand tools that are used to smooth metal, plastic and wood and to enlarge holes. **Abrasive papers** are also used to make metal, plastic and wood smooth. Sanding can be done by hand or by machine.

There is a wide range of files available. They are classified by their shape, the number of teeth per 25mm (**coarseness**) and how the teeth are arranged on the blade (their **cut**).

There are seven main shapes of files, and three main degrees of coarseness. Some files have a **safe edge**, which is an edge without teeth that can rub against material without filing it away.

The Safety Files

- Files must never be used without a handle. The tang can be very dangerous.
- Files should not be mistreated as they are brittle and break easily.
- Clean files regularly with a file card. This is like a wire scrubbing brush. Do not hit them on the workbench to clean them.

Rasps

Rasps are similar to files but they have individual teeth and remove material quickly. **Surform** files are a type of rasp. They are used on softer materials like wood, foam board and soft plastics.

Cross-Filing

This is the first stage of filing. The work piece should be securely clamped. The handle of the file in should be held in one hand and the tip in the other. A **cross-cut** file, has two sets of teeth crossing in a diamond pattern.

Filing should be in a straight line at 30° to the work piece, tip to handle. The pressure should be applied evenly along the whole length. The work piece should be checked regularly with a square edge

Draw Filing

This is the second stage of filing, when cross-filing marks are removed. A **single cut** file is used. This is a file with only one set of parallel teeth. It is drawn towards the user along the edge of the work piece. The pressure should be kept even along the whole length.

Abrasive Papers

When all the cross-filing marks have been removed, abrasive papers can be applied. The main types for wood are glass paper (buff), garnet paper (red) and aluminium oxide paper (green). **Emery cloth** and **silicon carbide** paper are used on metal. Silicon carbide paper is sometimes called 'wet and dry' paper.

Written Question

Spend about 8 minutes answering the following question. You will need some paper and something to write with.

Explain the main stages involved when finishing the edge of a piece of copper jewellery after it has been cross-filed.

(8 marks)

Finishing blocks

To be effective abrasive papers need to be wrapped around a wooden block.

Portable sanders

These are power tools that need to be used with care, and only under supervision. They are used for rapidly producing a fine finish over a large surface area.

Here's what you need to know...

about the various ways in which finishes can be added to resistant materials.

See *Design & Make It! Resistant Materials Technology* Revised pages 104–105 (96–97 earlier edition).

KEYWORDS

Do you know what the following terms mean?

- Finish
- Stains and varnishes
- Fixing and filling

Fine finishes

Most materials need to have some form of **finish** applied to them. This adds to their properties by making them stronger, water- or paint-resistant, more durable, more attractive, etc.

Sometimes materials need to be prepared first so that they will take the finish well. The surfaces of resistant materials can then be stained, polished or painted, and given a plastic-based finish to add an attractive, shiny and tough outer layer of protection.

Modern manufacturing technologies are providing an increasing range of **protective** and **decorative** finishes for resistant materials.

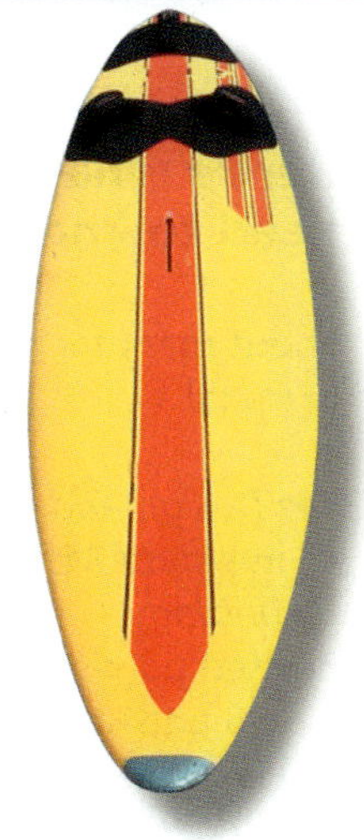

A surfboard needs a good water-resistant finish.

Preparation

Surfaces need to be clean and smooth. All marks or traces of adhesives need to be removed. Abrasive papers should be used as necessary in the order coarse to fine. Wood may need to be filled to cover screws or nails, or other indentations.

Stains and Varnishes

Stains can be applied to wood to change its colour, while still allowing the grain to show through. **Oils** and **polishes** can also be used on wood. Polyurethene **varnishes** provide an attractive and tough finish, and can be matt, satin or gloss. These all need to be applied using several thin coats that are rubbed down between applications.

Paint

Paint can provide a colourful and durable finish. There are two main types:

- **oil-based** which is tough, but more expensive. Polyurethane paints are particularly durable. Enamel paints are useful for detailed work, such as on small-scale product models.
- **water-based** (known as emulsion), and therefore not waterproof.

Fixing and Filling

When making presentation models from block modelling materials, small holes and gaps tend to occur where different materials are joined together. These need to be '**filled**' in some way, using body filler, plaster, etc. After they have been **sealed** they can be finished using acrylic, cellulose or water-based paints. Presentation models need to be very carefully finished if they are to look realistic.

Written question

Spend about 6 minutes answering the following question. You will need some paper and something to write with.

i) Explain briefly why most materials need to have some form of finish applied to them. *(2 marks)*

ii) Presentation models made from block modelling materials usually need to be filled. Explain what is meant by the term 'filled'. *(2 marks)*

iii) After being filled, presentation models are usually finished. Give two examples of suitable finishes that might be applied. *(2 marks)*

Here's what you need to know...

about manufacturing in metal.

See *Design and Make It! Resistant Materials Technology* Revised pages 58–59, 62–65 and130–131 (50–51, 54–57 and 122–123 earlier edition).

KEYWORDS

Do you know what the following terms mean?

- Forming
- Casting
- Cutting
- Joining
- Finishing

Manufacturing in Metal

Extracting metal

Metals are mined and then extracted from their mineral ores, usually by heating the ore until the metal melts. The metal is formed into bars, rods, sheets and tubes. **Iron**, **steel**, **copper** and **aluminium** are the four main metals used in manufacturing industry.

Forming

This means changing the shape of a piece of metal. These are common examples:

- **Press work** is used to produce thousands of products from coins to car bodies.
- **Extrusion** is when material is forced through a die to produce both solid and hollow shapes.
- **Drop forging**, or die forging, uses a die. A die is a tool or mould used to shape material that is usually pushed through or squeezed into it. Hot metal is squeezed between the two halves. One half is fixed to the anvil and the other to the hammer. The hammer is dropped on to the anvil and the two halves come together.
- **Turning** is for producing flat and cylindrical shapes. This is done on a centre lathe.

Casting (reforming)

This involves pouring or injecting molten metal into a mould and allowing it to cool. There are four main ways of casting.

- **Sand casting**. This is economic for making individual and small batches of items such as belt buckles, as labour and equipment costs are low.
- **Centrifugal casting**. This is an economic process for batches of less than 1000, as labour and material costs are quite high. Symmetrical objects such as pipes and boilers might be made this way.
- **Lost wax casting**. This process uses a pattern made from wax which 'disappears' when the molten metal is poured onto it. It is used for small finely detailed objects such as jewellery. It is economical and is used for batches of 100 to 5000.
- **Die casting** is used for high volume production, ie in excess of 1000 units, such as commemorative or collectable models.

Written Question

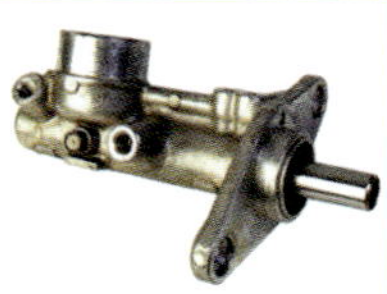

Spend about 6 minutes answering the following question.
You will need some paper and something to write with.

Here is a list of casting processes:

- Sand casting
- Die casting
- Centrifugal casting

Explain which of these casting processes would you use for the economic manufacture of the following products.

a) 50 engine cylinder blocks for a limited edition car.
b) 500 lamp posts.
c) 5,000 metal handles. *(6 marks)*

Cutting

Metal can be cut by:

- **Machining** using a cutting tool, e.g. a drill for holes, a milling machine for slots and other surface shapes and a lathe for cylindrical shapes and flat surfaces.
- **Thermal cutting** using gas or electric arc or a laser.
- **Hand tools**: e.g. hand saws, hand drills and chisels.

Joining

There are five main ways of joining metals:

- Mechanical fixings (e.g. rivets)
- Welding
- Soldering
- Brazing
- Adhesives

Finishing

There are many methods of finishing metals, including:

- Shot blasting
- Tumbling
- Anodising
- Pickling
- Electroplating
- Painting

Here's what you need to know...

about manufacturing in plastic (1).

See *Design and Make It! Resistant Materials Technology* Revised pages 34–35, 40–41 and 132–133 (26–27, 32–33 and 124–125 earlier edition).

KEYWORDS

Do you know what the following term means?

- Thermoplastics

WWW.

Go to: www.dtonline.org/areas/7/1/index.htm

Manufacturing in Plastic

Fantastic Plastic!

There are two main types of plastics, and several different ways of working with each type. This section is about forming **thermoplastics**. Section 3 is about forming thermosets, and cutting, joining and finishing both types of plastic.

Sources

Plastics are made from sustainable sources (e.g. trees: latex) and non-sustainable sources (e.g. crude oil: acrylic).

Thermoplastics

There are several different types of **thermoplastics**.

- **Acrylic** is used mainly in sheet form. It has a very hard shiny finish, comes in many bright colours and is easy to bend using heat. It can also be spun into threads and used for fabrics.
- **Polycarbonate** looks like acrylic but is extremely strong and is much more expensive. It is used for products such as police riot shields and helmets, and CDs.
- **Polythene** (polyethylene, etc.) is a lower cost plastic that is used for packaging, dustbins, seats and other low cost products. It cannot be glued but can be heat welded. It injection moulds well.
- **Polystyrene** is good for vacuum forming, but is not very strong and scratches easily. It comes in bright colours.
- **ABS** is used for high quality cases and many domestic products (walkmans, mobile phones, etc). It can withstand dropping without splitting. It injection moulds easily.
- **Nylon** is a very strong plastic that can be machined easily. It is sometimes used for simple bearings.

Blow moulding

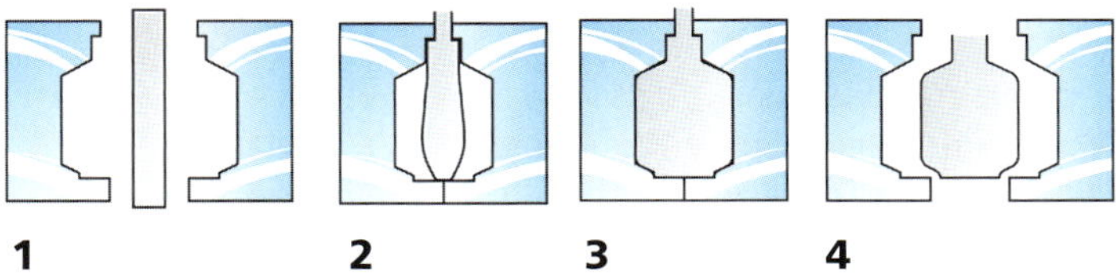

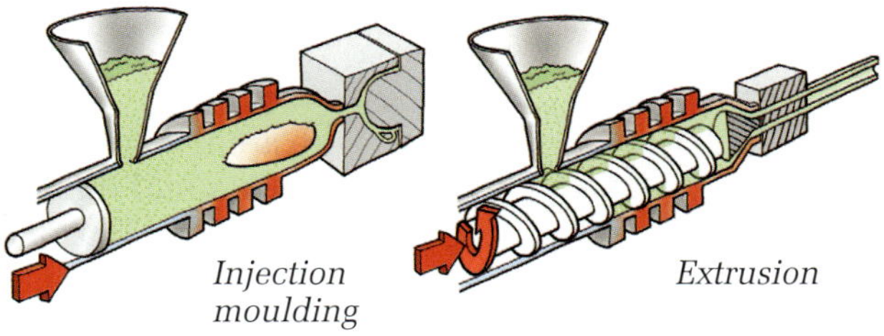

Injection moulding

Extrusion

There are several ways to form thermoplastics:

- **Strip heater** and bending jig: for line bending (see section 4 of this Topic).
- **Oven and former**: for forming double and single curvature forming.
- **Blow moulding**: for making dome shapes using hot thermoplastic sheet and compressed air. A restrictor can be used to provide a flat base or flutes like those on a drinks bottle
- **Vacuum forming**: for making complex shapes using hot thermoplastic sheet, a mould or pattern and evacuated air.
- **Extrusion**: for making tubes, rods and pipes using heated granules or powder fed through a die.
- **Injection moulding**: for making bowls, buckets, etc. using metal moulds and heated granules or powder.

Written Question

Spend about 8 minutes answering the following question. You will need some paper and something to write with.

Two methods of deforming plastic sheet are blow moulding and vacuum forming. Both use air pressure to change the shape of the plastic.

Explain the differences and similarities between these two methods.

(8 marks)

Here's what you need to know...

about manufacturing in plastic (2).

See *Design and Make It! Resistant Materials Technology* Revised pages 34–35, 40–41 and 132–133 (26–27, 32–33 and 124–125 earlier edition).

KEYWORDS

Do you know what the following terms mean?

- Thermosetting plastics
- Thermoplastics
- Cutting
- Joining
- Finishing

www.

Go to:
www.dtonline.org
www.plasticsnet.com

Even More Fantastic Plastic!

There are two main types of plastics, and several different ways of working with each type. This section is about forming **thermosets**, and cutting, joining and finishing both types of plastic. Section 2 is about forming thermoplastics.

Forming: Thermosets

The thermosets most commonly used in products are:

- **Polyester resin** is used for casting, embedding, encapsulation and Glass Reinforced Plastic (GRP) work (see below). A concave (or female) mould is coated with a releasing agent. Colour, if required, and then hardener are added to the polyester resin. The liquid resin is poured into the mould. This is like making a jelly in a mould.
- **Glass Reinforced Plastic** is used for making car and boat bodies and seats. Layers of glass fibre matting are placed over a former (a male mould). Polyester resin is applied between the layers of glass fibre matting to bond the material. When the resin cures or hardens, a rigid shell is produced. It can then be shaped and cut with hand or power tools.
- **Epoxy resin** is used as an adhesive, or to fill gaps.

Cutting

Both thermosetting plastics and **thermoplastics** can be cut by:

- **Machining** using a cutting tool, e.g. a drill for holes, a milling machine for slots and other surface shapes and a band saw for cutting flat shapes from a sheet. A lathe may also be used. (See section 4 of this Topic for more information.)
- **Hand tools**, such as junior hack saws, coping saws and files. (See also Topic 6 section 2.)

Joining

There are three main ways of joining plastics:

- **Mechanical fixings**: drill, tap and use of machine screws.
- **Adhesives**: Tensol cements are used for acrylic – these bond by melting the surfaces. Contact adhesives are used for bonding sheet thermoset plastics.
- **Welding**: using hot air.

Finishing

Thermoplastics, e.g. acrylic, do not need a protective coating. They are sold with a paper or plastic film covering that protects them from being scratched. To remove scratches, use **silicone carbide paper** with water then carefully machine buff with buffing compound and a mop. If pressed too hard the thermoplastic will melt. Make sure eye protection is used.

Written Question

Spend about 10 minutes answering the following question. You will need some paper and something to write with.

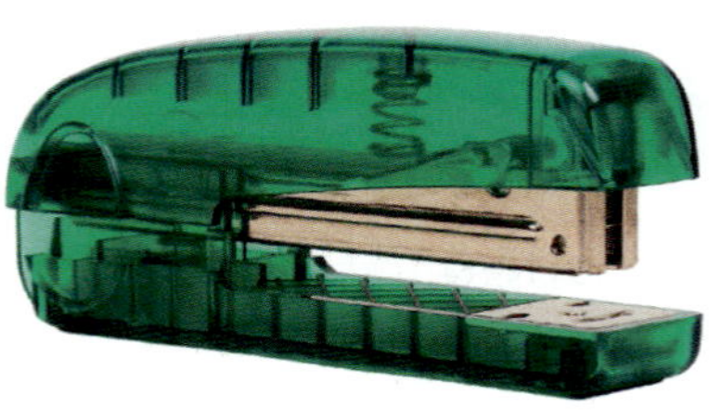

Polyester resin can be used in two different ways.

a) On the right is a photograph of a stapler that has a body made from plastic.
   i) Give a reason why it has been made from polyester resin *(1 mark)*
   ii) What are the four main manufacturing processes involved in making the plastic part of the stapler? *(4 marks)*

b) i) What do the letters GRP stand for? *(1 marks)*
   ii) What are the two main materials needed to create GRP? *(2 marks)*
   iii) Name two hand tools that could be used to cut and shape GRP once it has set. *(2 marks)*.

Here's what you need to know...

about manufacturing in wood.

See *Design and Make It! Resistant Materials Technology* Revised pages 100–105, 120–121 and 134–135 (92–97, 112–113 and 126–127 earlier edition).

KEYWORDS

Do you know what the following terms mean?

- Sawing
- Chiselling
- Planing
- Turning
- Joints
- Laminating
- Finishing

Wood you believe it!

There are a range of different forming, cutting, joining and finishing processes involved in working with wood and manufactured boards.

Sawing

There is a wide range of saws with different cutting properties and uses. (See also Topic 6, section 2.)

Chiselling

This is cutting with a wedge.

- **Bevel edged chisels** are used for removing small sections of wood. This is known as paring.
- **Firmer** and **mortise chisels** are used with a mallet for removing large quantities of wood. This is known as chopping.

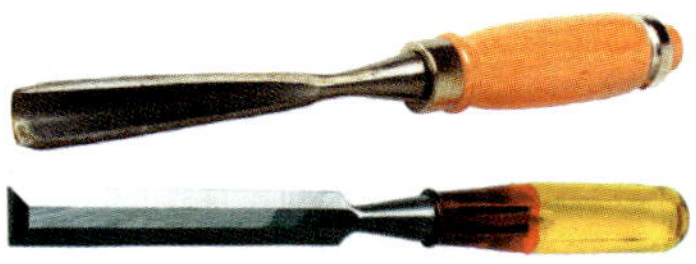

Planing

This is also cutting with a wedge.

- **Jack planes** are 350mm long and are used for planning wood flat and to size.
- **Smoothing planes** are 250mm long and lighter than Jack planes. They are used for finishing and planing end grain.

Turning

Special long handled chisels, **gouges** and **scrapers** are used to shape wood on a lathe. The wood is rotated either at the end of the lathe on a face plate, e.g. for making bowls, or between centres, e.g. to make chair legs.

Joints

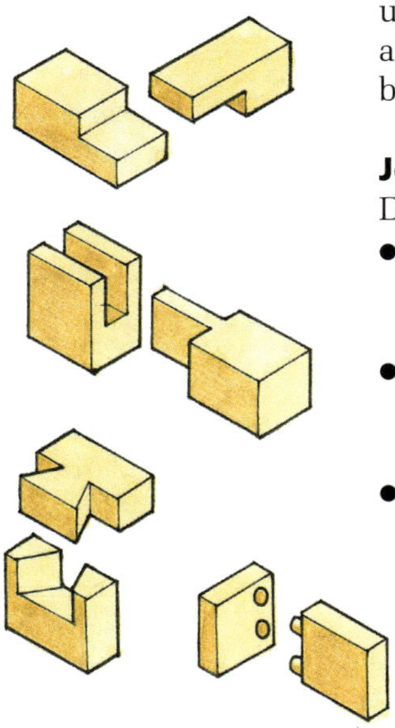

Different joints are used for different purposes.

- **Mitre** and **butt joints** are quite easy to make accurately, but they are not very strong. They are used for picture frames and other products that do not need to have strong joints.
- **Mortice** and **tenon joints** are very strong. They are used for furniture. They are difficult to make accurately as they need to be marked out and cut very carefully.
- **Dovetail** and **comb joints** are strong and decorative but need a lot of skill to make accurately. They are used on boxes and drawers.

Laminating

Thin sheets of wood are built up in layers with adhesive and bent to the required shape. **Formers** are used to support the curves until the adhesive dries. The results are very strong and used for tables, chairs and even bridges!

Finishing

The surface must be dust-free otherwise the finish will not be smooth as the dust will be trapped in the paint, stain or varnish when it dries. Cellulose sanding sealer is often used as it dries quickly and is easy to finish.

Written Question

Spend about 6 minutes answering the following question. You will need some paper and something to write with.

Using a sketch, explain the process of strengthening a simple butt joint between two pieces of 50mm square section wood. *(6 marks)*

Here's what you need to know...

about machine tools and techniques.

See *Design and Make It! Resistant Materials Technology* Revised pages 34–35, 65, 123 and 127 (26–27, 57, 115 and 119 earlier edition).

KEYWORDS

Do you know what the following terms mean?

- Pillar drill
- Machine saw
- Power saw
- Milling machine
- Centre lathe

WWW.

Go to: www.herts.ac.uk/lis/ltdu/projects/mm2

Power crazy

In industry materials are cut, formed and joined used **power-driven machine tools**. These are essential to ensure the accurate and rapid manufacture of large quantities of components. Smaller-scale versions of these machines are often found in designer's workshops for prototype and model-making.

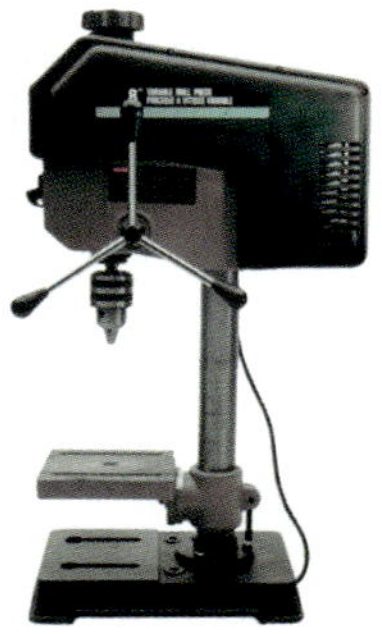

Pillar Drill: Wood, Metal and Plastic

A **pillar drill** is either mounted on the floor or bench. The chuck usually holds up to 13mm shank drills. It is driven by a belt drive that adjusts the speed. A fast speed should be used for small drills and soft materials. A slower speed should be used for large drills and harder materials.

The drill should be lowered slowly and carefully. The work piece must always be securely clamped, using a G clamp or machine vice, as work can 'snatch'. The height and angle of the table can be adjusted to the chuck. The chuck guard must always be used.

Machine and Power Saws

There are are a range of **machine** and **power saws**.

- **Circular saw**: fitted with rip or cross cut blades for wood.
- **Powered hacksaw**: for cutting steel up to 150mm diameter.
- **Band saw**: different blades for wood, metal or plastic. The blade is a continuous band (hence the name) running round wheels in the machine.
- **Electric jig saw**: uses a variety of blades for wood, metal or plastic. It has reciprocating action and adjustable speed.

Milling Machine: Metal and Plastic

Milling machines have rotating cutters. They are used to cut horizontal and vertical surfaces and slots. There are two types of milling machine: **horizontal** and **vertical**, depending on where the cutters are mounted. Horizontal machines use cutters that are wheel shaped, vertical machines use cutters which look like drills. The table moves in all three axes (x, y and z). Work must be securely clamped and the guard used.

Centre Lathe: Metal, Plastic and Wood

This machine cuts very accurately and uses tools that are held in a tool post that can be moved in two directions. The work piece is held and rotated while the chosen cutting tool on the tool post cuts the material. Internal holes can be bored into bar using drills or a boring tool.

Written Question

Spend about 6 minutes answering the following question. You will need some paper and something to write with.

Outline how you would modify the processes of drilling an 8mm hole using a pillar drill, in a 100mm x 100mm piece of each of the following materials:

a) Mild steel. *(2 marks)*
b) Redwood. *(2 marks)*
c) Acrylic. *(2 marks)*

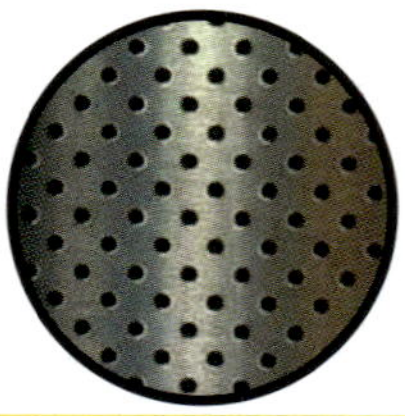

Wood Turning Lathe

The work piece is held on a face plate, in a chuck or between centres depending on the shape to be cut (wide and short, thin or long). The tools are held by hand on a tool steady. It is important to wear goggles. Finishing work by sanding requires the tool steady to be removed for safety.

Strip Heater: Plastic

This is basically a long thin heating element held in a case. It is used for heating thermoplastic sheet so that it can be bent. It is a wise precaution to wear leather gloves to hold the material. Very short pieces of plastic can be held with pliers or a small clamp. (See also Section 2 in this topic.)

Vacuum Former: Plastic

The machine is used to heat thermoplastic sheet so that it can be sucked over a mould or former to create a three dimensional shape. (See also Section 2 in this topic.)

Here's what you need to know...

about production process systems.

See *Design & Make It! Resistant Materials Technology* Revised pages 126–129 (118–121 earlier edition).

KEYWORDS

Do you know what the following terms mean?

- Production processes
- One-off production
- Batch production
- Mass production
- Continuous production

Production, Production, Production

Different products are manufactured in different ways, depending on how many are needed. There are four types of **production process**:

- one-off production.
- mass production.
- batch production.
- continuous production.

One-Off Production

One-off production is when a craftsperson produces a single item to a very high standard. It is expensive in labour costs and materials. An example of a one-off production system would be the method a cabinet maker would use to make a single specialist item.

A one-off product can be made by more than one person.

Batch Production

This is where a fixed number of identical products are made using simple jigs, templates, moulds or other aids to speed up production and control repeat quality. It can be a large or a small batch. Tasks and equipment are shared and manufacturers can quickly respond to changes in the market, using machine tools to produce a batch of different products, for example, kitchenware. A batch of one particular colour or shape is made at a time.

Batch production can be done by one person, or by many people.

Mass Production

Mass production is where workers on a production line quickly make identical products all day, every day. '**Tooling up**' – setting up the production line machinery – can be very expensive, but once enough products have been sold to recover the cost, the cost of manufacture is low. Changing the production line to respond to changes in the market can be expensive and take a long time. One example is making cars.

Many mass-production lines are now highly **automated**, with only a small number of workers involved.

Continuous Production

This is where the same product is made continuously, round the clock. This type of production is used where stopping and re-starting the production process takes a long time and is very expensive. An example is making steel in strip and sheet form. The heating systems involved are expensive and difficult to stop and restart quickly.

Written question

Spend about 6 minutes answering the following question. You will need some paper and something to write with.

State which production system is used for manufacturing:

a) a family car
b) a high performance racing bike

Explain the differences between the production systems used for each. *(6 marks)*

Here's what you need to know...

about structural systems.

KEYWORDS

Do you know what the following terms mean?

- Frame structure
- Shell structure
- Static load
- Dynamic load
- Compression
- Tension
- Torsion
- Shear
- Bending

Supporting Structures

Structures are all around us. You are probably sitting on one and are inside one! Structures are **manufactured** and **natural**: a tree and a bridge are both structures. A structure needs to support both itself and the load or force put on it – a car needs to support itself and the people it carries, for example.

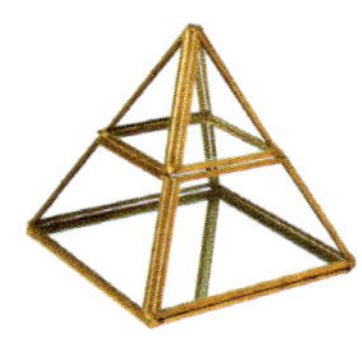

There are two main types of structural systems: frame structures and shell structures.

- **Frame structures** have a frame as the main part of the structure. They are made up of several parts or members. Triangulation is used to stabilise them.
- **Shell structures** are made from a continuous skin and it is their shape that provides the strength.

Many structures are a combination of frame and shell structures

Stabilising Structures

When a structure is stable, it is in equilibrium. It is able to support itself and the load that it is carrying, e.g. a bridge can support itself and the traffic that passes over it. When the bridge collapses it is called **structural failure**. Engineers have to calculate all the forces that will act on a bridge and then add an additional safety margin.

Loads

There are two types of load: static loads and dynamic loads.

- **Static loads** do not move, such as a can of paint on a shelf.
- **Dynamic loads** move: such as a child on a trampoline. These produce greater forces than static loads.

Forces

There are five main types of force:

Compression: e.g. the legs on a chair are compressed by the person sitting on it.

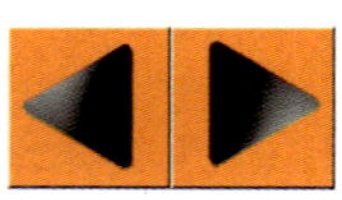

Tension: e.g. the chains on a hanging basket are held in tension by the weight of the basket.

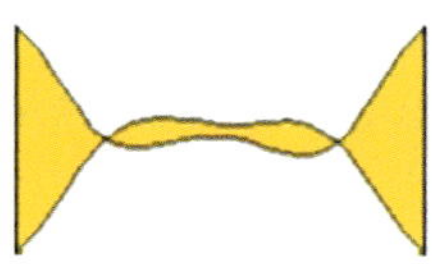

Torsion: e.g. the elastic band on a model aircraft is twisted (torsion) by turning the propeller.

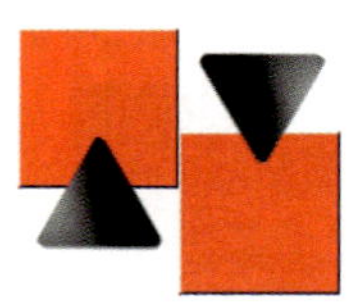

Shear: e.g. when cutting hair with scissors, each hair has the two blades acting on it in different directions.

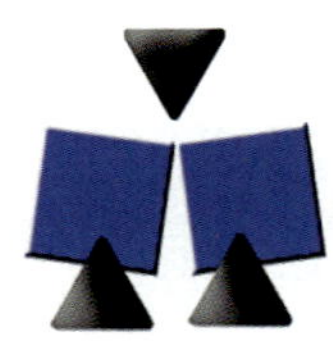

Bending: the combination of tension on the lower and compression on the top surfaces.

Written Question

Spend about 5 minutes answering the following question. You will need some paper and something to write with.

Look carefully at the photographs of the hanging basket and bracket and the man on the step ladder. State whether the forces acting on the objects at points **a**, **b**, **c**, **d** and **e** are each either under compression or in tension. *(5 marks)*

Here's what you need to know...

about mechanical systems.

See *Design and Make It! Resistant Materials Technology* Revised pages 84–87 and 90 (76–79 and 82 earlier edition).

KEYWORDS

Do you know what the following terms mean?

- Linear motion
- Reciprocating motion
- Rotary motion
- Oscillating motion
- Levers
- Cams
- Cranks
- Sliders
- Gears
- Gear train

www.
Go to:
www.flying-pig.co.uk/mechanisms/index.html
www.dtonline.org/apps/menu/app

Move It!

Mechanical systems help make machines more efficient by reducing the amount of energy required to use them. They can also change one type of motion into another. Levers, cams, cranks and sliders use four different types of motion.

- **Linear motion** is movement in a straight line.
- **Reciprocating motion** is a repeated backwards and forwards movement in a straight line.
- **Rotary motion** is circular movement.
- **Oscillating motion** is a repeated left to right/right to left movement that follows a curved path.

First class

Second class

Third class

Levers

There are three classes of **lever**; first, second and third. All levers consist of a **load**, a **fulcrum** (or pivot) and an **effort**. These can be in different positions. The position of the load, fulcrum and effort decides which class a lever belongs to.

Cams

Cams are irregular shaped wheels, or can be round with off-centre holes. They rotate around an axle and have a follower resting against them.

Cranks and Sliders

Crank and **slider** mechanisms can be used to change reciprocating motion to rotary motion and to change rotary motion to reciprocating motion.

Getting into gear

Gears are mechanisms based on the principle of the toothed **wheel and axle**. The driver drives the driven gear. When two gears are meshed, they turn in opposite directions. To turn them in the same direction, use an **idler** gear between them. A small gear meshed to a large gear decreases speed. A large gear meshed to small gear increases speed.

When two or more gears are meshed together it becomes a **gear train**. A compound gear train is made from pairs of meshed gears. They are for making large speed changes or getting different outputs moving at different speeds.

Pulleys

Pulleys can make it easier to lift loads by altering the mechanical advantage. A pulley increases the amount of effort that can be put into lifting the load. Pulley systems can also move a load in a different direction to that in which you exert a force. There are a number of different types of pulley systems, for example:

- Single fixed pulley
- Block and tackle
- Belt drives

Written Question

Spend about 11 minutes answering the following question. You will need some paper and something to write with.

i) Name the four basic types of motion, explain how each one moves, and give an example of its application. *(8 marks)*

ii) On the right is a photograph of a wheelbarrow. Identify the marked positions of the load, effort and fulcrum marked at **a**, **b** and **c**. *(3 marks)*

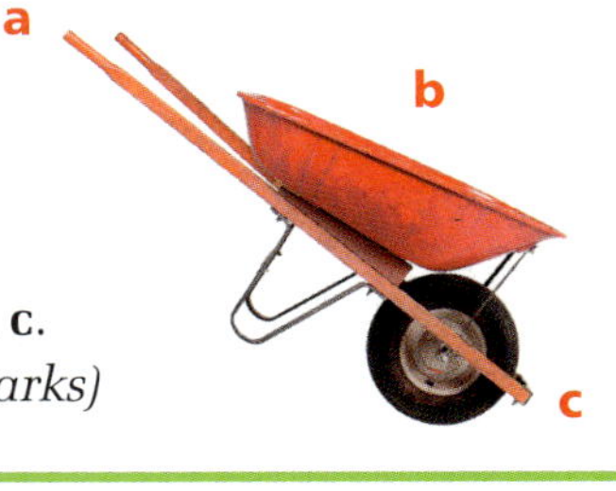

Here's what you need to know...

about how industrial production systems are organised.

See *Design and Make It! Resistant Materials Technology* Revised pages 124–125 (116–117 earlier edition).

KEYWORDS

Do you know what the following terms mean?

- Production system
- Sub-system
- Sub-assembly

Working the System

A **production system** is a sequence of **processes**, **materials** and **components**. These need to work together to produce a product in the most efficient way. If the system is not efficient, time, materials and components may be wasted, and the manufacturer's profit will be reduced.

The production manager finds ways of improving the production system, to make the product more efficiently.

In production

Systems are made up of inputs, transformation processes and outputs.

In a factory:

- the **inputs** include raw materials, components, employees.
- the **transformation** processes include the manufacturing process, the packaging process and quality checks.
- the **outputs** include the product, paid employees and waste materials.

Input: raw materials

Transformation: production processes

Output: finished products in the warehouse

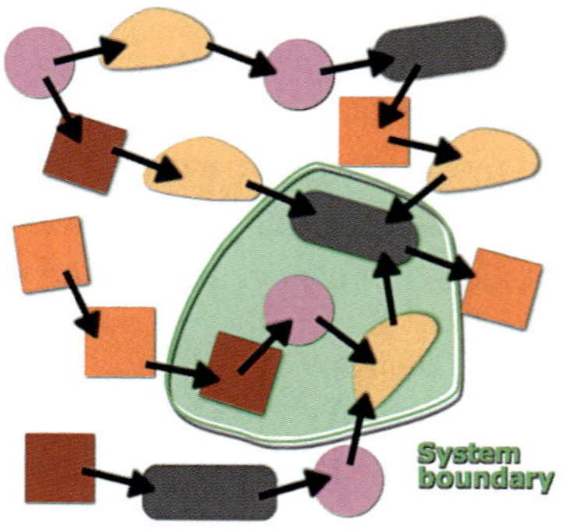

Sub systems

All systems are part of other, larger systems. The **system boundaries** have to be defined before you can study it. In a sweet factory you might study the system from raw materials arriving to packaged sweets leaving, but you might just look at how the finished sweets are wrapped and packaged. This is a **sub-system**.

The transformation process is a series of sub-systems. For example, when a production line is being planned, the different operations are identified. This may include **sub-assemblies**, where components are assembled in manufacturing cells before joining the production line.

Staying in control

Flexible manufacturing systems (FMS) help keep things running smoothly, and ensure the most efficient use of time and energy and cost-effectivess. These systems are highly adaptable and can include CNC lathes and milling machines, robotic arms and automatic guided vehicles (AGV) transport systems.

The **JIT (Just in Time)** production control system ensures that materials and components arrive when they are needed. **Gantt charts** provide a complex and accurate production schedule.

Written Question

Spend about 9 minutes answering the following question. You will need some paper and something to write with.

Give an account of the inputs, transformation processes and outputs that you might find in a factory making flat-pack furniture.

(9 marks)

Here's what you need to know...

about quality control systems.

See *Design and Make It! Resistant Materials Technology* Revised pages 124–125, 128–129 and 136–139 (116–117, 120–121 and 128-131 earlier edition).

KEYWORDS

Do you know what the following terms mean?

- Tolerance
- Quality Control
- Quality Assurance

Quality counts

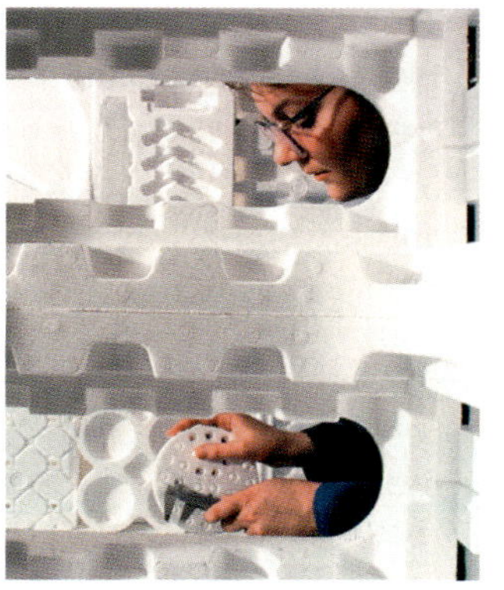

The need for consistent quality in products is very important. A company needs to make sure it does not develop a reputation for manufacturing products that are **unreliable**. It also wants to minimise the costs of **rejects** and **repairs**.

How accurate is it?

Gauges, jigs, templates and measuring systems (e.g. callipers, micrometers, electronic sensors, etc.) are used to check accuracy.

Can you tolerate it?

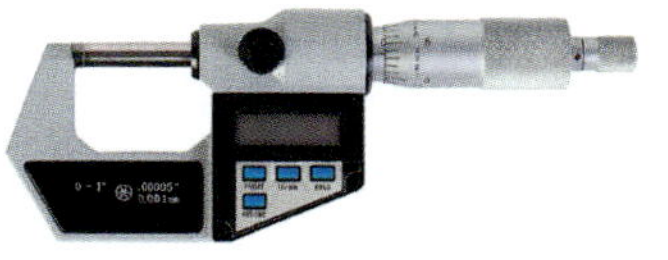

Tolerance is the allowable variation in size from the original design.

In a complex product, such as a high performance engine, the tolerance level for each component will be small. The more accurately a product is made, the better the quality will be.

In simpler products, for example, a soup can, the tolerance is greater. This allows for a wider range of acceptable sizes for the tin can. For example, the can could be 70mm ± 2, in diameter. ± means 'plus or minus'. So a tolerance of 70mm ± 2 diameter means that it can be anywhere in the range 68 to 72mm. Any larger or smaller and the lid could not be fitted.

Quality Control

Quality control involves a series of methods used to check products as they are made. In most cases it would take too long to inspect every item on a production line. For this reason a **sample** is examined – one in every hundred, or thousand perhaps, or once a minute, or maybe half-hour, depending on the product.

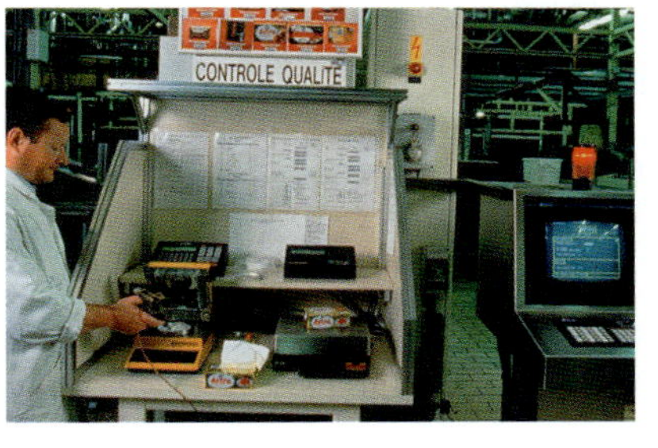

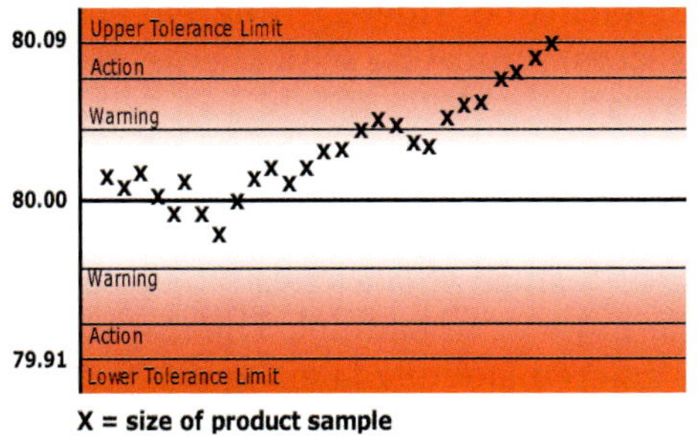

X = size of product sample

The results of the test are recorded and compared: if they become regularly unsatisfactory then the production process is stopped. This **feedback** about the state of the production system helps identify problems before too many **sub-standard** products are produced.

Some quality control checks will be undertaken and recorded by production line workers. Computer-controlled systems are frequently used for machine processes. Computers can supply feedback quickly, they can process large amounts of information at high speed and keep logs of data over long periods of time that can be easily accessed and analysed.

Written Question

Spend about 6 minutes answering the following question. You will need some paper and something to write with.

Explain the difference between quality assurance and quality control. *(6 marks)*

Quality Assurance

Quality Assurance is the overall approach that a company takes to keeping standards high in all aspects of its procedures, documentation, communications and product quality. Quality assurance provides a guarantee to consumers that all its products and services are of a certain standard.

Here's what you need to know...

about using ICT in Resistant Materials Technology.

See *Design & Make It! Resistant Materials Technology* Revised pages 10–11, 137 (129 earlier edition).

KEYWORDS

Do you know what the following terms mean?

- Word processing
- DTP (Desk-Top Publishing)
- Internet
- Search engine
- Database
- Spreadsheet

www.

Go to:
www.microsoft.com/education/?ID=Tutorials

Using ICT

ICT is used all through the process of design. Some types of software are intended particularly for designers to use. These are discussed in sections 3 and 4. However, many other types of software are also used to research, plan and communicate design ideas.

Word Processing / DTP

Computer programs that deal with text are used to prepare design briefs, specifications, and research and development reports. **Word processors** are fine if little visual material (e.g. photographs, drawings, charts, diagrams) is to be included. If images are going to be included then a **DTP (Desk-Top Publishing)** program is usually better, as this will make sure that the text and pictures are accurately placed on the page.

On the WWW

There is an extraordinary amount of information available on the **Internet**. This can be very useful when a designer needs to find out more about the product being developed. As well as reference text, the Internet is also a good source of images. A **search engine** is used to help find information that will be relevant. The Internet can also be used to send **emails** to specific people or organisations to request or discuss information.

Databases

Databases are organised information about a particular subject. Databases can be electronically searched in different ways to sort out the information required, such as suitable materials, target markets, anthropometric data, etc.

Some databases exist on CD-Roms, or on an organisation's own network. Others can be accessed over the www.

Spreadsheets

Spreadsheets can be used as complex calculators. When data is entered (e.g. costs, weights, sizes, etc.) into a spreadsheet, the software can perform complex mathematical tasks on the data quickly and reliably. The data can also be automatically presented as a graph or chart, which is easier to understand than large sets of numbers.

Designers use spreadsheets for entering research data, considering different costing options, creating parts lists and materials lists, communicating with clients, etc.

Written Question

Spend about 6 minutes answering the following question. You will need some paper and something to write with.

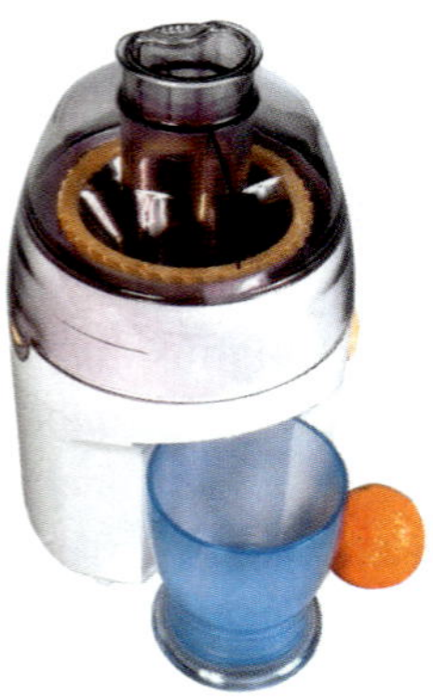

Give three examples of software packages, other than CAD/CAM, that a designer might use when developing ideas for a new food processor. Explain briefly what each could be used for.

(6 marks)

Planning

There are a number of programs that can be used to plan projects where a range of tasks need to be carried out in a particular order or in parallel. The stages and time available is entered and displayed **graphically** as a diagram or flow chart. The impact of changing one stage of the process can be quickly realised on the rest of the plan.

Presentation

Designers often use **presentation** programs to communicate their ideas to clients. These are easy to set up and use. On-screen 'slides' present the information clearly to everyone in the room. Key points, charts and diagrams, photographs, etc. can all be put into the slides for the presentation.

Here's what you need to know...

about how CAD/CAM systems are set up for use.

See *Design & Make It! Resistant Materials Technology* Revised page 20

KEYWORDS

Do you know what the following terms mean?

- CAD
- CAM
- CAD/CAM
- Inputs
- Transform
- Outputs
- CNC

www.

Go to:

www.tep.org.uk

www.cadinschools.org/home/

Uses of ICT

In the design of products made from resistant materials, ICT is often used to produce things like:

- working drawings,
- models, prototype drawings and renderings to show a client or customer
- patterns and moulds for product manufacture.

CAD stands for Computer Aided Design. **CAM** stands for Computer Aided Manufacture. In a **CAD/CAM** system, inputs are transformed into outputs.

What You Put In

Inputs include things such as text, data, photographs, graphic shapes and 3D forms. These are entered into the computer by means of keyboards (plus mouse), scanners and graphic tablets.

The inputs are **transformed** electronically within the computer. That is to say they can be changed as a result of instructions given by the user, using a computer program.

What You Get Out

The **outputs** are often in the form of text, data or images. These might be printed in black and white or colour and produced on a laser or inkjet printer, or a plotter.

The outputs can also be in the form of information sent to another computer controlled device, such as a **CNC** (Computer Numerically Controlled) milling machine, router or laser cutter to produce the design in 3D.

CNC Milling Machines

Milling machines are mainly used for cutting harder materials like metals. There are three different types of milling machines that can be controlled by a computer.

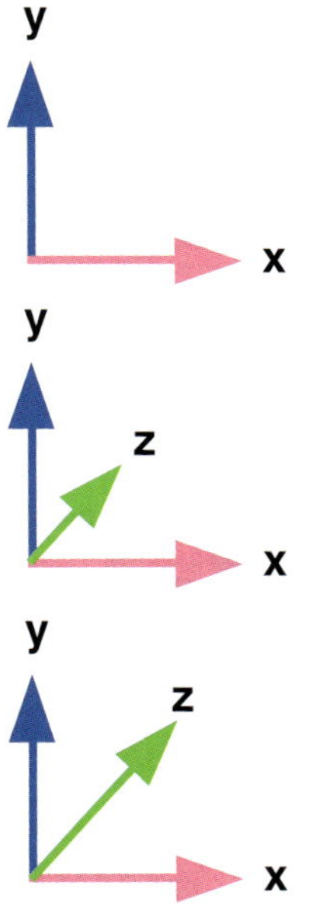

- A **2 axis miller** is a profile cutter. This means it can only cut out a two-dimensional shape. One axis is called X and the second is called Y.
- A **2.5 axis miller** has an X and a Y axis, and a small Z axis. This means it can also do engraving and simple milling of sheet material.
- A **3 axis miller** has fully controllable X, Y and Z axis adjustment allowing true three-dimensional cutting.

A **CNC router** is just like the milling machine, but has a very high speed motor and is used mainly for machining plastics and wood. Like the 3 axis miller it can cut along 3 axes. A **CNC laser cutter** is also used for machining plastics and wood and can also cut along 3 axes.

CIM

CIM (Computer Integrated Manufacture) is a totally integrated system of manufacture. Design, machine programming and operation, quality control, stock control and movement, costings, etc., are all carried out using a single networked system rather than a series of individual CAD/CAM computers.

Written Question

Spend about 6 minutes answering the following question. You will need some paper and something to write with.

A designer is developing a new wooden toy in which a series of cut-out parts interlock together.

State one input device, one type of computer program and one output device that might be used, and say what it would be used for.

i) a) Input device
   b) Purpose
ii) a) Computer program
   b) Purpose
iii) a) Output device
   b) Purpose

(6 marks)

Using 2D and 3D CAD

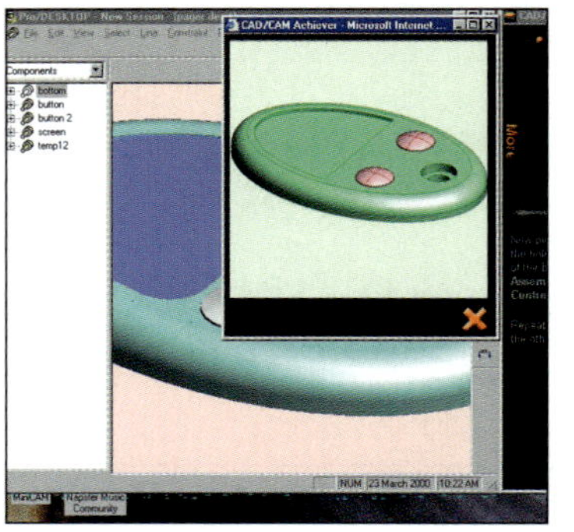

Here's what you need to know...

about using 2D and 3D CAD programs.

See *Design & Make It! Resistant Materials Technology* Revised pages 10–11, 20, 39, 97 and 118–119 (89 and 110–111 earlier edition).

KEYWORDS

Do you know what the following terms mean?

- 2D CAD
- 3D CAD
- Three-dimensional representation
- Wire-frame model
- Surface model
- Virtual prototype
- Solid model
- Rapid prototype
- PCB

2D CAD

2D CAD is used mainly for developing things that will be printed or published on the web. Sometimes, however, resistant material product designers use 2D CAD packages to develop surface patterns or finishes for 3D products.

3D CAD

3D CAD programs make it possible to create a **three-dimensional representation** of a product on-screen. The image can be rotated in any direction, zoomed in and out, or 'walked around' to give an impression of what it would look like in reality.

CAD makes it easy to generate, develop, modify and enhance ideas. New possibilities can be created and evaluated in minutes, reducing the need to keep making more and more real prototypes or models.

Modelling Methods

There are three main techniques involved:

- **Wire-frame modelling**, in which the object is seen on screen as a series of lines, or 'contours'. These models are not very realistic to look at but are quicker to modify, manipulate and rotate.
- **Surface modelling**, where colour, shading and texture are added to a wire-frame model to give a more realistic impression of what the object would look like. Different lighting sources can be set to highlight different parts of the product. These pictures can be printed out to create a '**virtual prototype**' to show to a client for approval, or for use in promotional material.
- **Solid modelling**, in which the form shown on screen can be mathematically analysed by the computer to provide information on changes in the object's mass, volume, etc.

Rapid Prototypes

From these on-screen models it is possible to produce rapid prototypes. **Rapid prototypes** are quick and cheap to produce. The modelling data can be sent directly from the computer to a **CAM** machine that will make a **3D model** out of a solid block, or cut sections or layered profiles, which are built up rather like a contour map.

Drafting Programs

Once a design has been finalised, accurate technical drawings and plans can be created on screen using special '**drafting**' programs. As with a conventional technical drawing, plan and elevation views (3rd angle orthographic projections) are used. These can be created from 3D modelling software like **Pro-Desktop** where the product is drawn in 3D from the start. Some drafting software works the other way where 3D images of parts are created from 2D views.

Once checked and approved the data from these drawings can be sent directly to automated machinery to make the parts. Using the Internet to transfer the data means the machine that actually makes the product can be anywhere in the world.

PCBs

Special software can be used to design electronic printed circuit boards. These can be printed out and used directly as masks to produce printed circuit boards (**PCBs**).

Written Question

Spend about 6 minutes answering the following question. You will need some paper and something to write with.

A team of designers is creating the casing for a new mobile phone. Name three applications of 3D CAD/CAM that could be suitably used in the design development process. For each, briefly explain their use and benefit. *(6 marks)*

When to use CAD/CAM

Advantages of using CAD/CAM

There are a number of important advantages and disadvantages in using **2D** and **3D CAD/CAM**. The main advantages are:

- **speed**, particularly in terms of experimentation and communication
- **accuracy** (i.e. fewer mistakes), especially when things are copied.
- **easy sharing** of data, storage of data and access to data

All these significantly help increase productivity and reduce production costs and ensure that products are made to a consistently high standard of performance and appearance.

Here's what you need to know...

about the advantages and disadvantages of the industrial use of CAD/CAM.

See *Design & Make It! Resistant Materials Technology* Revised pages 10–11, 20, 39, 97, 118–119 and 136–137 (89, 110–111 and 128–129 earlier edition).

KEYWORDS

Do you know what the following terms mean?

- 2D CAD
- 3D CAD
- CAD/CAM
- Global markets

WWW.

Go to: www.cadinschools.org/home/

Disadvantages of using CAD/CAM

There are also a number of disadvantages:

- computer technology is expensive to buy and maintain.
- using computer technology requires a lot of training and experience.
- using computer technology means less evaluation of real materials and 3D forms.
- it is easy to lose data if it is not regularly 'backed up'.

The Impact of ICT

ICT is having a significant impact on the way designers design, and how products are made and sold. It continues to lead extensive changes in the workplace, the high street and the home:

- New products can be created, modelled and changed far more quickly.
- Increased automation has produced higher quality products, but less work for the unskilled.
- Robotic devices can work in conditions that are unsafe for humans.
- It is possible to access information and communicate with others much more quickly and easily, opening up **global markets**.

Written Question

Spend about 9 minutes answering the following question. You will need some paper and something to write with.

i) Give three examples when you used, or could have used, ICT in your Resistant Materials Technology course. Name the type of programs you used and say briefly what you did. *(6 marks)*

ii) State two advantages and one disadvantage of ICT over using other methods when designing and making resistant materials products in school. *(3 marks)*

Here's what you need to know...

about the different roles of the client, the designer, the manufacturer and the user.

See *Design & Make It! Resistant Materials Technology* Revised pages 8–9

KEYWORDS

Do you know what the following terms mean?

- Client
- Manufacturer
- User/Consumer
- Retailer

www.

Go to:
www.technology.org.uk/
www.baddesigns.com/index.shtml

The Client, the Designer, the Manufacturer and the User

New products only happen as a result of the collaboration of a client, a designer (or team of designers), a manufacturer and the potential users.

A great idea for a new product might be worthless unless:

- the **client** believes that it will sell well.
- the **manufacturer** can understand how it needs to be made, including how many are required.
- the **consumer** realises the features and benefits it has, and wants to purchase it.

The Client

The client is basically the person, company or organisation that puts up the money to pay for the design, production and promotion of a new product. If it is successful they will make a lot of profit. If it fails to sell they will lose a lot of money. The client, therefore, needs to be confident that the features, appearance and selling price of the product will appeal to the current demands of potential customers. The client usually sets the **initial brief** for the designer to work to.

In some situations the client may also be the manufacturer, or the designer may work directly for the client or the manufacturer.

The Manufacturer

Manufacturers want to know about the materials, sizes, components, time limits and manufacturing processes involved. They need to be sure they can make a profit on the manufacturing process, taking into account the costs of materials and labour, machine tools, factory space, storage, transportation, energy, insurance, etc.

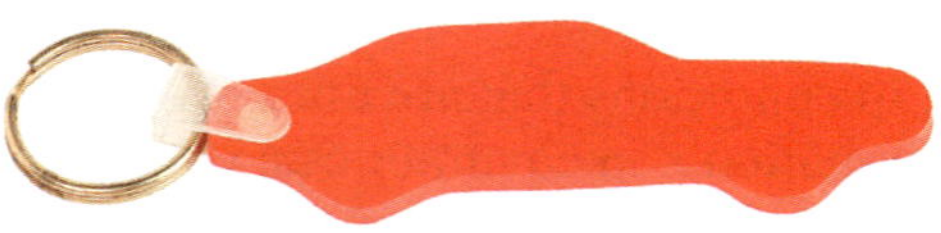

The User

The **user** wants to know what the product does, what benefits it will bring them, how much it costs and where they can buy it. They expect a new product to do what it is supposed to do, be safe and last a reasonable length of time in use.

The Retailer

In most cases the user will purchase a product from a **retailer**. The retailer will need to be sure that a product they have in stock is likely to sell well and not just sit on a shelf unsold, taking up valuable display and storage space.

Written Question

Spend about 6 minutes answering the following question. You will need some paper and something to write with.

A design team has developed a new flat-pack computer workstation. They now need to present their ideas to:

i) the client (a large retail furniture chain)
ii) a manufacturer.
iii) potential customers.

For each of the above suggest a suitable use of ICT and briefly explain what information it would help communicate to each. *(6 marks)*

Here's what you need to know...

about industrial production planning and costing.

See *Design & Make It! Resistant Materials Technology* Revised pages 22–23 and 126–129 (14–15 and 118–121 earlier edition).

KEYWORDS

Do you know what the following terms mean?

- Flow chart
- Inputs
- Outputs
- Feedback

www.
Go to:
www.technology.org.uk/

Industrial Production Planning

Making a product in large quantities in a factory is a very different process from making just one or two in a small workshop. The production line needs to aim to:

- minimise and automate the number of construction processes.
- reduce waste.
- increase speed.
- ensure quality is consistent.

Often the design of a product is developed, or needs to be changed, to help achieve this. ICT and other new automated technologies often involve higher initial investment, but significantly reduce costs in the long run.

Basically, the more identical products that are manufactured, the cheaper it becomes in terms of the purchase of materials and components, and the use of labour, energy and factory space. However, high production volumes bring an increased risk of unsold products, and a lack of choice for the consumer.

The final manufacturing process needs to be fully documented in order to ensure all the workers know what they are supposed to be doing and when, without the designer being there to tell them.

Flow Charts

A **system** can be drawn up using a **flow chart**. This identifies the **inputs** and **outputs** and shows the different stages of a sequence of events. It can also show where **feedback** occurs. Different symbols are used to show different parts of the process. The symbols are linked together by arrows. These show how one event follows on from another. The aim is to make the flow chart as simple to understand and follow as possible.

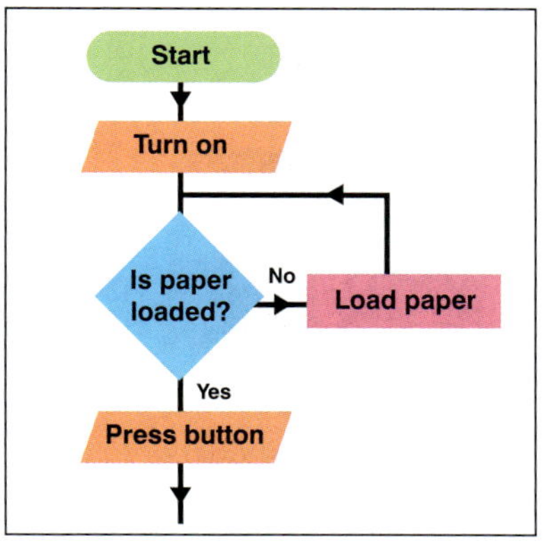

These symbols are usually used when writing computer programs.

Note how the movement of a component is shown by an arrow symbol.

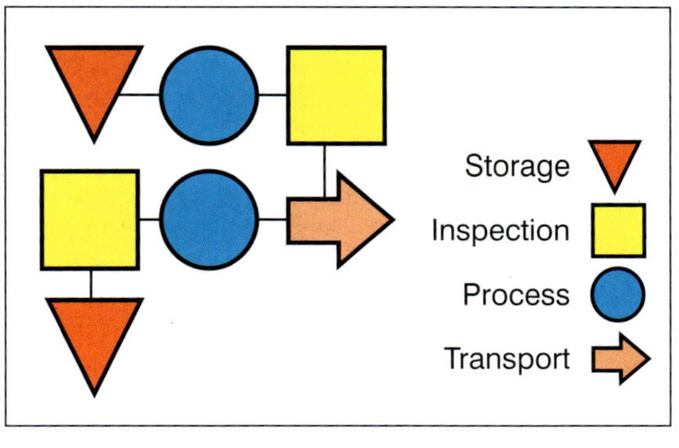

In Production

When planning a **production line**, a slightly different set of symbols is used. In a complicated product, each stage of manufacture would have its own flow chart, with a master chart showing how they all flow together.

Each product will need its own **production chart**, often according to whether it will be made in small or large batches, or on a mass production line. These methods may require very different tools, machines and storage facilities.

Written Question

Spend about 6 minutes answering the following question. You will need some paper and something to write with.

A designer is planning to produce a batch of 500 high quality wooden CD storage racks. Listed below are 12 stages of the production process.

Place these stages into a flow chart using appropriate production symbols.

(6 marks)

Start
Prepare the wood
Move the wood to workstation 1
Cut the pieces to the correct length
Check the sizes are correct
Assemble the sides
Move the assembly to workstation 2
Add the pre-made metal dividers
Add self-adhesive feet
Check quality of construction
Move to storage for packaging
End

Marketing Matters

Here's what you need to know...

about advertising and marketing products to potential customers.

KEYWORDS

Do you know what the following term means?

- Marketing

www.

Go to:
www.mgmcommunications.com
www.marketresearch.org.uk/mrindustry/faqs.htm
www.mori.com

Marketing involves a lot more than asking people to fill in a questionnaire, or designing clever advertisements to sell people things they don't really want. Successful businesses invest a great deal of time and money in marketing. They consider what are known as the '**4 Ps of Marketing**':

- the **Product** – what goods or services do people need or want?
- the **Place** – where do people want to be able to purchase the product from?
- the **Price** – how much are people willing to pay for the product?
- the **Promotion** – how can people best be informed that the product is available and how it can be purchased?

A lot of research is needed before, during and after the process of product development to make sure all the money invested in its production will not be wasted. There's no point making a product that no one wants to buy, or that people don't know exists.

The Product

Not everyone wants the same thing. Different products appeal to different people. Most products are designed for particular groups of people (a market). Companies look for a '**gap in the market**' to make a product for, i.e. a product that is not currently available aimed at a particular market that will want it. For example, a sportswear company might identify a market for a range of lower-cost and lower-performance shoes for a particular sporting activity that young children are becoming interested in.

The Place

It is important to make sure that it is easy for a customer to buy a new product. There is a range of **retail outlets** that can be considered for distribution, e.g.:

- local shops.
- supermarkets.
- shopping centres.
- mail order catalogues.
- the Internet.

The choice of outlet needs to be carefully matched to how long the target market is willing to wait and how far they are willing to travel. We expect to be able to buy a toothbrush in a specialist local shop like a chemist on the day we realise we need a new one, even though we know it might be cheaper in a supermarket where there might be more choice. However, we're prepared to travel to a large shopping centre to choose a new sofa, and wait several weeks for the particular colour we want to be delivered. As a general rule, consumers are prepared to wait longer for higher cost items, which are more individual.

The Price

Getting the price right is essential. A high price suggests high quality materials and exclusivity, but restricts the number of people who will be able to afford to purchase it. A low price can indicate something that is not likely to perform well or last long, so may also put people off.

Some products are deliberately designed for these extremes.

Most products are priced somewhere in the middle. It is essential to establish the **target price range** before design work can begin.

The Promotion

Advertising is crucial, particularly when a new product is launched. If people don't know the product is available they can't be expected to buy it! When they become aware of a new product they then need to be informed about what it does (its **features**) and why they might need or want it (its **benefits**).

Advertising is expensive so it needs to be targeted at the people most likely to want the product. This is known as **target marketing**. The advertisements will be placed where the target market is most likely to see them, for example, in specialist magazines, during TV programmes they are most likely to watch, at places and events they are most likely to attend, alongside other products they are likely to purchase.

Written Question

Spend about 10 minutes answering the following question. You will need some paper and something to write with.

You have been asked to suggest an appropriate method of advertising a new bar of soap aimed at male and female users in the 15 to 25 age range. The product is part of an 'environmentally friendly' product range. This is the main benefit of the soap.

Suggest:

i) A suitable magazine in which to place a full-page colour advertisement.

ii) A suitable TV programme to show a 15-second advertisement.

iii) A suitable celebrity to endorse the product.

iv) A suitable location and/or event to present the product.

v) Two suitable descriptive words that might be used in an advertisement.

Give a reason for each of your choices. *(10 marks)*

Here's what you need to know...

about the use of quality assurance symbols.

See *Design & Make It! Resistant Materials Technology* Revised page 92 (84 earlier edition).

KEYWORDS

Do you know what the following terms mean?

- Quality control
- Quality assurance

www.

Go to: www.bsi-global.com/Education/index.xalter

Guaranteeing Quality

Consumers want to be sure that a product will work properly and safely. The law protects consumers' rights too. That is why it is very important for manufacturers to control quality and let consumers know that quality standards have been met.

A **performance specification** will tell the manufacturer what the product should be able to do. Manufacturers test a sample from each production run against the specification to make sure the product meets the standards required of it. This process for checking quality against the specification is called **quality control**.

Manufacturers will often give a warranty or guarantee on a product to show that it meets its standards. This is **quality assurance**: letting consumers know that quality standards have been met.

Making a Product Safe to Use

Safety is an important part of quality assurance. People want to know that the things they buy are safe to use. A manufacturer uses quality control to check the product meets its safety requirements, and then uses quality assurance methods to let consumers know these checks have been done and passed.

A range of **visual symbols** has been designed to help consumers quickly identify that the product they are buying conforms to the necessary safety standards. Some relate to **BSI (British Safety Institute) Standards**, others to European Standards.

Here are some common examples:

The Kite Mark

This symbol confirms that the product complies with the requirements of the BSI. The product needs to be independently tested, and must be re-tested at regular intervals.

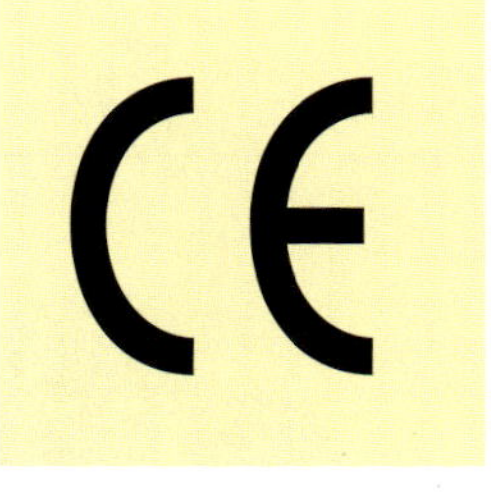

The CE Mark

The letters CE show that a manufacturer claims that the product meets the essential safety requirements of the European Commission.

The e Mark

This indicates that the contents of a package contain, on average, the quantity stated. The mark must be at least 3mm high and placed near to the quantity details.

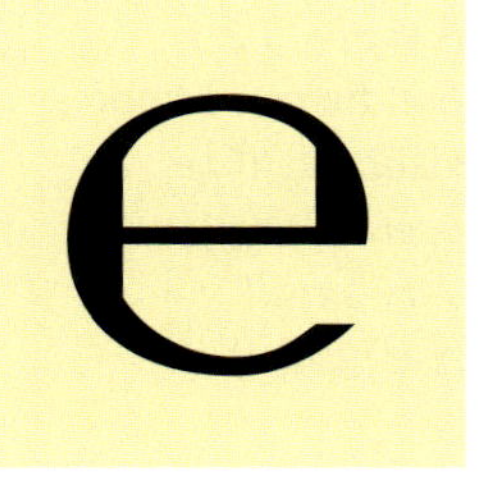

Age Warning

This symbol must appear on a toy (or its packaging) not suitable for children under 3 years old. The warning may be given by written words instead.

Written Question

Spend about 4 minutes answering the following question. You will need some paper and something to write with.

The kite mark and CE mark are common examples of safety standard symbols. Give a brief explanation of the different meanings they have. *(4 marks)*

Here's what you need to know...

about drawing conventions used in industry.

See *Design & Make It! Resistant Materials Technology* Revised pages 38–39 (30–31, earlier edition).

KEYWORDS

Do you know what the following terms mean?

- Third angle orthographic projection
- Parts list
- British Standards
- Scale

Orthographic Projection

Working drawings are intended to enable someone other than the designer to make a product. For this reason they have to be clear and accurate. Various conventions and common symbols are used to help achieve this.

In a working drawing 3D products are shown as a series of **plans** and **elevations**. The positioning on the drawing of these views follows a particular arrangement known as **third angle orthographic projection**. The plan is placed immediately above the front elevation. The side elevation is lined up to the left or right of the front elevation.

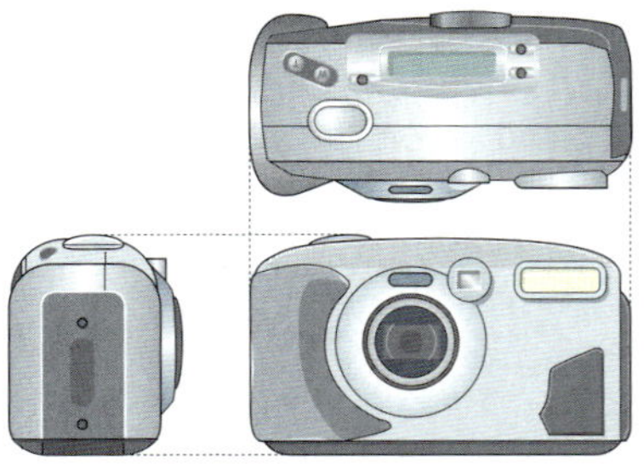

Written information in the form of a table, or '**parts list**', is provided to specify part numbers, materials and finishes and assembly instructions.

British Standards

The overall dimensions of the object or structures are drawn in a particular way, following a set of **British Standard** conventions (PD7308). Briefly:

- dimension lines are placed to the side of the drawings.
- projection lines indicate the parts of the drawing the dimensions relate to.
- projection lines should not touch the drawing of the object or structure.
- dimension and projection lines should be half the thickness of the lines of the object or structure.

Note that some CAD drafting programs use slightly different conventions, but you are expected to use British Standard conventions in the written paper.

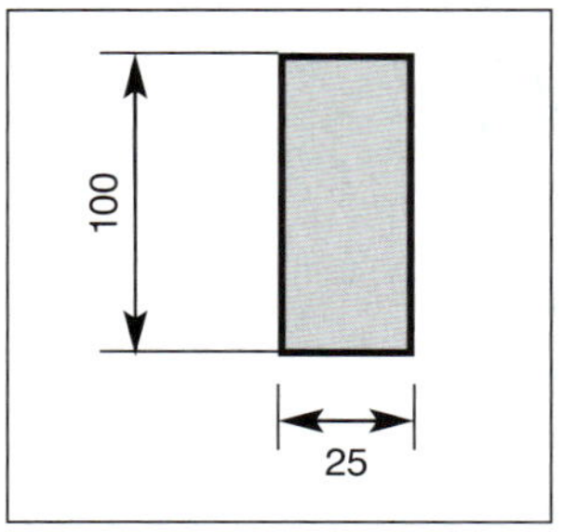

There are other drawing conventions for working drawings, such as for drawing cross-sections, cut-aways and exploded views.

Scale

Many drawings are drawn to **scale**, that is to say at a different size to the real object or structure.

A scale of 2:1 (i.e. twice normal size) might be used to draw a very small component.

In a drawing at a scale of 1:10, the drawing is one-tenth of the real, full size.

A scale of 1:500 might be used to draw a large public building.

Written Question

Spend about 10 minutes answering the following question. You will need some paper and something to write with.

Draw a quick, freehand, orthographic drawing of the outline of the audiocassette box below.

i) Show a front elevation/view, an end elevation/view and plan. *(3 marks)*

ii) Show three major dimensions to BS conventions. *(3 marks)*

iii) Show the BS symbol for orthographic projection. *(2 marks)*

iv) State the scale used. *(2 marks)*

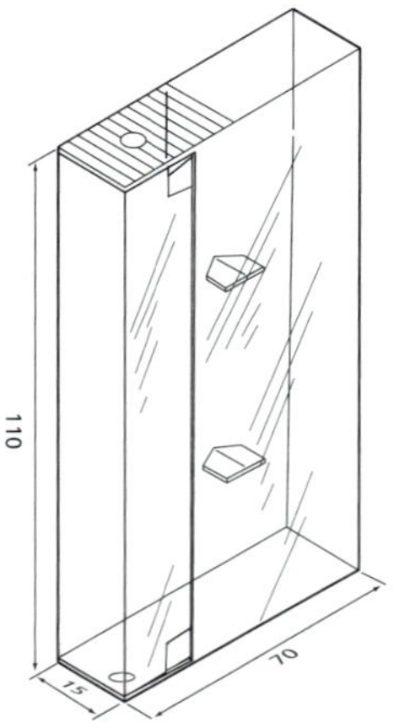